High-Pressure Physical and Chemical Behaviors of Minerals and Rocks

High-Pressure Physical and Chemical Behaviors of Minerals and Rocks

Editors

Lidong Dai
Haiying Hu
Jianjun Jiang

MDPI • Basel • Beijing • Wuhan • Barcelona • Belgrade • Manchester • Tokyo • Cluj • Tianjin

Editors
Lidong Dai
Chinese Academy of Sciences
China

Haiying Hu
Chinese Academy of Sciences
China

Jianjun Jiang
Chinese Academy of Sciences
China

Editorial Office
MDPI
St. Alban-Anlage 66
4052 Basel, Switzerland

This is a reprint of articles from the Special Issue published online in the open access journal *Minerals* (ISSN 2075-163X) (available at: https://www.mdpi.com/journal/minerals/special_issues/8th_FAES).

For citation purposes, cite each article independently as indicated on the article page online and as indicated below:

LastName, A.A.; LastName, B.B.; LastName, C.C. Article Title. *Journal Name* **Year**, *Volume Number*, Page Range.

ISBN 978-3-0365-7290-1 (Hbk)
ISBN 978-3-0365-7291-8 (PDF)

Cover image courtesy of Lidong Dai

Contents

About the Editors

Lidong Dai

Dr. Lidong Dai has been a professor at the Key Laboratory of High-temperature and High-pressure Study of the Earth's Interior, Institute of Geochemistry, Chinese Academy of Sciences (China) since 2001. As an associate director of the lab, he was elected as part of the Hundred Talents Plan from the Chinese Academy of Sciences on December, 2013, and as well as part of the Ten Thousand Talents Plan of the Central Organization Department from People's Republic of China on December, 2021. As a visiting researcher and postdoctoral fellowship, he has conducted more than 5 years of substantial academic communications and collaborations with the Department of Earth and Planetary Sciences, Yale University, USA; Mineral Physics Institute, Department of Earth and Planetary Sciences, New York State University, USA; and the Department of Earth and Planetary Sciences, Tokyo Institute of Technology, Japan. He has published more than 150 papers in journals indexed in the Web of Science, such as *Earth and Planetary Science Letters, Journal of Geophysical Research: Solid Earth, Physical Review B, Inorganic Chemistry,* and *Applied Physics Letters.* His research is mainly focused on high-pressure mineral physics and high-pressure material physical chemistry by virtue of multianvil high-pressure apparatus, diamond anvil cells and shock waves. At present, he is focused on high-pressure Raman spectroscopy, high-pressure Fourier transform infrared spectroscopy, high-pressure electrical conductivity and high-pressure first-principles theoretical calculations in conjunction with the recovered high-resolution transmission electron microscopy and atomic force microscopy of some representative minerals and rocks in the upper mantle, mantle transition zone, and subduction zone of deep Earth interiors, as well as many crucial functional materials.

Haiying Hu

Dr. Haiying Hu is an associate professor at the Laboratory of High-temperature and High-pressure Study of the Earth's Interior, Institute of Geochemistry, Chinese Academy of Sciences (China) since 2012. As a visiting researcher, she has conducted more than 2 years of academic communications and collaborations with the Department of Earth and Planetary Sciences, Yale University, USA and the Mineral Physics Institute, Department of Earth and Planetary Sciences, New York State University, USA. She has published more than 20 papers in journals indexed in the Web of Science, such as *Earth and Planetary Science Letters, Journal of Geophysical Research: Solid Earth,* and *Gondwana Research.* Her research mainly focuses on the physical (electrical, phase transition) and chemical properties of geomaterials, including minerals, rocks, fluid, melting at high temperature and high pressure the by the use of in situ impedance spectrum measurement in combination of confocal Raman spectroscopy, Fourier transform infrared spectroscopy, X-ray diffraction, and electron microprobe analysis, in order to provide significant constraints on Earth structure, composition, physical state, and process.

Jianjun Jiang

Dr. Jianjun Jiang received his PhD degree in December 2016. After that, he worked as an assistant professor at the Key Laboratory of High-temperature and High-pressure Study of the Earth's Interior, Institute of Geochemistry, Chinese Academy of Sciences (China). During 2007 to 2020, as a postdoctoral researcher and visiting scholar, he served in the Institute for Planetary Materials, Okayama University (Japan) and the Department of Geosciences, Princeton University (USA). His main research direction focused on the electrical conductivity of mineral and rock using multianvil press and Brillouin scattering elastic wave velocity using diamond anvil cells. He published more than 10 papers in journals indexed in the Web of Science. Sadly, Dr. Jianjun Jiang passed away in August 2021. This Special Issue commemorates his significant contributions to Professor Lidong Dai's research team.

minerals

MDPI

Editorial

Editorial for Special Issue "High-Pressure Physical and Chemical Behaviors of Minerals and Rocks"

Lidong Dai * and Haiying Hu *

Key Laboratory of High-Temperature and High-Pressure Study of the Earth's Interior, Institute of Geochemistry, Chinese Academy of Sciences, Guiyang 550002, China
* Correspondence: dailidong@vip.gyig.ac.cn (L.D.); huhaiying@vip.gyig.ac.cn (H.H.)

The eighth "From Atom to Earth" symposium on high-pressure science and earth science was held at the Key Laboratory of High-temperature and High-pressure Study of the Earth's Interior, Institute of Geochemistry, Chinese Academy of Sciences (IGCAS), the People's Republic of China, from 2 to 5 July 2021. As the conference president, Professor Lidong Dai from the IGCAS Key Laboratory of High-temperature and High-pressure Study of the Earth's Interior was also the conference organizer. Participants from Chinese top colleges and research institutions gathered to communicate some of the most recent progress in the field of the physical and chemical behaviors of minerals and rocks under high-temperature and high-pressure conditions.

This research topic of the high-pressure physical and chemical behaviors of minerals and rocks, as hosted by *Minerals*, sets its root in the eighth "From Atom to Earth" symposium on high-pressure science and earth science in order to provide an opportunity to explicitly display new developments in high-pressure mineral physics. This Special Issue it contains ten original articles and one review paper, which mainly focus on high-pressure experimental studies and the theoretical calculation of results on minerals and rocks in high-temperature and high-pressure conditions.

As we known, the electrical conductivity of minerals and rocks (e.g., olivine, clinopyroxene, wadsleyite, epidote, amphibole, peridotite, etc.) in the deep Earth interior are very sensitive to several factors, including temperature, pressure, water content, titanium content, dehydration effect, oxidation−dehydrogenation effect, crystallographic orientation, structural phase transition, and iron content [1–5]. In their review article, Hu et al. [6] retrospectively examined experimental measurements on the high-temperature and high-pressure electrical conductivity of hydrous minerals and rocks in the deep Earth crust, upper mantle, and subduction zone by means of the multi-anvil high-pressure apparatus and diamond anvil cell. Using the electrochemical alternating impedance spectroscopy method, some recent research progress for several recently reported electrical conductivity on four representative hydrous silicate minerals, including the lower case titanium-bearing and water-bearing synthetic polycrystalline olivine aggregates, single crystal epidote, single crystal amphibole, and as well as polycrystalline kaolinite powder, was summarized and outlined by them under conditions of high temperatures and pressures. Specially noted, five main influential factors, including the titanium content, water content, dehydration effect, oxidation−dehydrogenation effect, and structural phase transition on the high-pressure electrical conductivity for those of hydrous silicate minerals, were deeply explored. Furthermore, some comprehensive remarks on several possible future research aspects have been discussed in detail.

A series of new high-pressure experimental results are here presented for many typical minerals and rocks by means of several different high-pressure equipment, such as autoclave, multi-anvil press, diamond anvil cell, shock wave, etc. In order to measure in situ rock porosity under high-temperature and high-pressure conditions and different isostatic pressure environments, Liu [7] designed and applied an experimental assemblage

Citation: Dai, L.; Hu, H. Editorial for Special Issue "High-Pressure Physical and Chemical Behaviors of Minerals and Rocks". *Minerals* **2023**, *13*, 477. https://doi.org/10.3390/min13040477

Received: 17 March 2023
Revised: 21 March 2023
Accepted: 25 March 2023
Published: 28 March 2023

on the base of the high-pressure apparatus of autoclave, located at the Key Laboratory of High-temperature and High-pressure Study of the Earth's Interior, Institute of Geochemistry, Chinese Academy of Sciences, Guiyang, China. In order to test the accuracy and validity of the high-temperature and high-pressure measurement assembly, the effect of experimental parameters, including the initial gas pressure at the inlet, the time needed for the gas to reach equilibrium, and as well as the time needed for vacuuming on the porosity experiment, was fully examined under different isostatic pressures. The cylindrical 304 stainless steel, cylindrical copper samples, and cylindrical sandstone under the high isostatic pressure of up to 200 MPa, as three dominant test samples, the authors believe that it is quantitatively verified at the high interrelation between the degree of rock porosity and its correspondent isostatic pressure. By virtue of YJ-3000t multi-anvil high-pressure apparatus and a Solartron-1260 impedance spectroscopy analyzer, Wang et al. [8] measured the electrical conductivities of the dry hot-pressed sintering gabbro with various mineralogical proportions ($Cpx_X Pl_{100-X}$, X = 0, 10, 20, 30, 40, 50, 60, 70, 80, 90, and 100 vol% (the signals of *Cpx* and *Pl* stand for the single crystals of clinopyroxene and plagioclase, respectively) at temperatures of 773–1073 K and pressures of 1.0–3.0 GPa. The functional relationship between the electrical conductivity of the sample and influential factors (e.g., temperature, pressure, and mineralogical proportion) was deeply established. Furthermore, laboratory-based electrical conductivity–depth profiles for the hot-pressed sintering gabbro with various mineralogical proportions and temperature gradients were successfully constructed. They concluded that, although their acquired electrical conductivity results on the dry hot-pressed sintering gabbro with various mineralogical proportions cannot explain the high conductivity anomaly in the oceanic crust and West African craton, it can provide one reasonable constraint on the mineralogical composition in those of representative gabbro-rich regions. Ren [9] performed the high-pressure phase transition from the α–quartz powder to coesite using YJ-3000t multi-anvil high-pressure apparatus at a temperature range from 500 °C to 700 °C, confining pressures between 1.5 GPa and 1.8 GPa and differential stress conditions. The occurrence of quartz–coesite phase transition was observed under conditions of the specific temperature of 500 °C and high differential stress with a relatively longer reaction time. He believes that the appearance of a recovered coesite phase is highly related to differential stress, reaction time, and reaction temperature, with coesite formation being a multifactorial coupling process. By the hydrothermal diamond anvil cell, Chen et al. [10] investigated the variations in the Raman spectra of pyrite under conditions of temperature ranges from 113 K to 853 K and atmospheric pressure, and as well as temperature ranges from 297 K to 513 K and pressures up to 1.9 GPa. They found that the high-pressure hematite phase began to form at the temperature point of 653 K and, subsequently, the starting material of the pyrite phase was thoroughly transformed into the hematite phase at the temperature point of 688 K, and, finally, the hematite phase would be melted at a temperature higher than 853 K. The formation of the high-pressure phase of pyrite occurred at an increase in temperature at each correspondent initial pressure (group 1: 0.5 GPa, group 2: 1.1 GPa, group 3: 1.7 GPa, group 4: 1.9 GPa), which showed no evidence for the occurrence of a chemical reaction or sample decomposition. Pressure and temperature effects are evident for groups 1 and 2, whereas, for groups 3 and 4, the temperature displayed a relatively larger effect than pressure and resulted in a sharp decrease in A_g and E_g modes. Zhang et al. [11] performed in situ Raman scattering and electrical conductivity experiments on calcite in order to study the phase state and structural stability of the sample during both compressed and decompressed processes using the piston-cylinder and four column-type diamond anvil cells under conditions of temperature ranges from 298 K to 873 K and pressure ranges from 1 atm to 19.7 GPa. They observed the pressure hysteresis and reversible behavior during the process of structural phase transformations from $CaCO_3$-I to $CaCO_3$-II phases, from $CaCO_3$-II to $CaCO_3$-III phases, and as well as from $CaCO_3$-III to $CaCO_3$-VI phases at a room temperature condition. At three representative pressure points (i.e., 10.5 GPa, 12.5 GPa, and 13.8 GPa), the phase transition temperatures between $CaCO_3$-III and $CaCO_3$-VI phases were individually determined. Furthermore, the pressure–temperature diagram in the phase boundary of calcite was well established at temperature ranges from 298 K to 773 K, pressures up to 16.0 GPa, and a depth up to 480 km, and further, its potential geophysical implications were discussed in detail. Gao et al. [12] performed a series of investigations via high-pressure Raman spectroscopy using the *BX90*-type diamond anvil cell and theoretical calculations, utilizing the density functional theory (DFT), on the structural phase transition from zircon to reidite at temperature ranges from 298 K to 800 K and pressure ranges from 0.6 GPa to 26.0 GPa. Their experimental results indicate that the high-pressure reidite phase becomes thermodynamically more stable in comparison with the starting material of the zircon phase at conditions of the fixed pressure of 8 GPa and atmospheric temperature, and the slope of the phase boundary within the relatively lower temperature ranges of 298–800 K abruptly differs from those of previously reported results at the relatively higher temperature ranges of 1100–1900 K. Further, in comprehensive considerations of their experimentally observed equilibrium phase boundary and theoretical calculations based on the density functional theory (DFT) method, it indicates that the kinetic effect of the zircon–reidite phase transition is obvious, and there exists a sufficiently large energy driving force provided by an overpressure to overcome the activation energy barrier below a critical temperature of approximately 880 K. Sun et al. [13] adopted the laser indirectly driven shock compression for quartz to investigate the sound velocity of quartz at extremely high-pressure ranges from 270 GPa to 870 GPa during the lateral unloadings at the Shenguang-III prototype laser facility (SG-IIIp), located at the Laser Fusion Research Center, Chinese Academy of Engineering Physics, Mianyang, China. On the base of those of obtained shock wave data, they also calculated all of these parameters' values, including the high-pressure shock wave velocity, density, and Grüneisen parameter from sound velocity data. All of these obtained experimental results on the sound velocity of quartz under the laser indirectly driven shock compression can provide a new insight to support the dissociation and metallization for the liquid quartz at extremely high-pressure conditions.

On the other hand, some recently acquired results of theoretical calculations including all of these physical characteristics, including the self-consistent thermal expansion, heat capacity, Grüneisen parameters, equation of state, compressibility, and vibrational properties for the minerals in the regions of the upper mantle and subduction zone at conditions of a high temperature and high pressure are displayed in detail in this Special Issue. Using an iterative numerical approach, Su et al. [14] theoretically calculated a series of self-consistent thermodynamic parameters, including the thermal expansion, heat capacity, and Grüneisen parameters on diopside within relatively wider temperature ranges from 0 K to 2000 K and pressure ranges from 0 GPa to 20 GPa. According to all of these obtained thermodynamic parameters' data on the diopside of major mantle silicate mineral, they extrapolated the adiabatic temperature gradient and geotherm of an eclogitic upper mantle. Furthermore, their theoretical calculating results can reveal that the adiabatic temperature gradient of the eclogitic upper mantle model is lower than the pyrolite model at the same depth, which can result in a relatively slower increase in temperature than that of a pyrolitic upper mantle and lead to a similar adiabatic geotherm for both pyrolitic and eclogitic upper mantle within depth ranges from 200 km to 410 km. By virtue of the atomistic computer simulations with the modified ClayFF classical force field method, Tararushkin [15] theoretically calculated the equation of the state, compressibility, and vibrational properties on the brucite of hydrous hydroxide mineral in the subducting zone at conditions of temperatures up to 873 K and pressures up to 15 GPa. Their theoretically acquired pressure and temperature dependencies of the crystalline lattice constant, unit cell volume, compressibility, bulk elastic modulus, and its corresponding high-order derivative, thermal expansion coefficient, the structural M–O–H bond angle, the structural hydroxyl bond length, the angular distribution of the hydroxyl orientation along the {001} crystallographic plane and spectra of vibrational density of states on brucite were comprehensively compared with a diverse set of available experimental data including X-ray diffraction, neutron scattering, infrared spectroscopy, and Raman spectroscopy. They demonstrate that

ClayFF-MOH, as simple and approximate as it is, can be quite accurate in predicting the physical characterization of brucite, which greatly expands the area of its applicability to the simulations of other hydrous minerals with a much more complex crystalline structure under the subduction zone conditions.

In addition, Liu et al. [16] investigated the geological characteristics of Lvwen stone with the yellow-green carbonate jade gemstone using conventional gemological testing methods and analytical techniques, including X-ray diffraction, Fourier transform infrared spectroscopy, ultraviolet–visible spectroscopy, laser ablation plasma mass spectrometry, and scanning electron microscopy (SEM). They found that the chemical composition of Lvwen stone is mainly composed of Ca element, with lesser amounts of Mg, Mn, Cu, Zn, Fe, and other trace elements with the electronic transition absorption band of iron (III) and copper (II) intra-ions; the mineralogical composition of Lvwen stone is calcite element, and trace-element- and crystal-size-induced colors result in its characteristic of having a banded appearance. They indicate that both the electronic transitions of Fe^{3+} and Cu^{2+} intra-ions give rise to the unique yellow-green color of the material. Lvwen stone is produced by ultra-high-pressure tectonic fluids in a relatively closed, reducing environment, and the green matrix was formed earlier than the white bands.

Author Contributions: L.D. designed the project. H.H. and L.D. wrote the initial draft of the work. L.D. and H.H. corrected and recognized the final paper. All authors have read and agreed to the published version of the manuscript.

Conflicts of Interest: The authors declare no conflict of interest.

References

1. Dai, L.; Liu, X.; Manthilake, G.; Saltas, V.; Hu, H. *High-Pressure Physical Behavior of Minerals and Rocks: Mineralogy, Petrology and Geochemistry*; Frontiers Media SA: Lausanne, Switzerland, 2023; pp. 1–179. [CrossRef]
2. Dai, L.; Hu, H.; He, Y.; Sun, W. Some new progress in the experimental measurements on electrical property of main minerals in the upper mantle at high temperatures and high pressures. In *Mineralogy*; René, M., Ed.; IntechOpen: London, UK, 2022; Chapter 2, pp. 15–38. [CrossRef]
3. Dai, L.; Karato, S. High and highly anisotropic electrical conductivity of the asthenosphere due to hydrogen diffusion in olivine. *Earth Planet. Sci. Lett.* **2014**, *408*, 79–86. [CrossRef]
4. Dai, L.; Karato, S. Electrical conductivity of wadsleyite at high temperatures and high pressures. *Earth Planet. Sci. Lett.* **2009**, *287*, 277–283. [CrossRef]
5. Hu, H.; Dai, L.; Li, H.; Sun, W.; Li, B. Effect of dehydrogenation on the electrical conductivity of Fe-bearing amphibole and its implications for the high conductivity anomalies in subduction zones and continental crust. *Earth Planet. Sci. Lett.* **2018**, *498*, 27–37. [CrossRef]
6. Hu, H.; Dai, L.; Sun, W.; Zhuang, Y.; Liu, K.; Yang, L.; Pu, C.; Hong, M.; Wang, M.; Hu, Z.; et al. Some Remarks on the Electrical Conductivity of Hydrous Silicate Minerals in the Earth Crust, Upper Mantle and Subduction Zone at High Temperatures and High Pressures. *Minerals* **2022**, *12*, 161. [CrossRef]
7. Liu, L.; Li, H.; Zhou, H.; Lin, S.; Li, S. Design and Application of a Rock Porosity Measurement Apparatus under High Isostatic Pressure. *Minerals* **2022**, *12*, 127. [CrossRef]
8. Wang, M.; Dai, L.; Hu, H.; Sun, W.; Hu, Z.; Jing, C. Effect of Different Mineralogical Proportions on the Electrical Conductivity of Dry Hot-Pressed Sintering Gabbro at High Temperatures and Pressures. *Minerals* **2022**, *12*, 336. [CrossRef]
9. Ren, D. Study on the Phase Transition from Quartz to Coesite under High Temperature and High Pressure. *Minerals* **2022**, *12*, 963. [CrossRef]
10. Chen, J.; Li, H.; Yuan, Y.; Zhang, M.; Shuai, S.; Wan, J. Raman Spectroscopic Studies of Pyrite at High Pressure and High Temperature. *Minerals* **2022**, *12*, 332. [CrossRef]
11. Zhang, X.; Dai, L.; Hu, H.; Li, C. Pressure-Induced Reverse Structural Transition of Calcite at Temperatures up to 873 K and Pressures up to 19.7 GPa. *Minerals* **2023**, *13*, 188. [CrossRef]
12. Gao, Y.; Zheng, Z.; Zhao, X.; Liu, Y.; Chen, J.; Li, Y.; Xiong, M.; Zu, X.; Mei, S. In Situ Raman Spectroscopy and DFT Studies of the Phase Transition from Zircon to Reidite at High P–T Conditions. *Minerals* **2022**, *12*, 1618. [CrossRef]
13. Sun, L.; Zhang, H.; Guan, Z.; Yang, W.; Zhang, Y.; Sekine, T.; Duan, X.; Wang, Z.; Yang, J. Sound Velocity Measurement of Shock-Compressed Quartz at Extreme Conditions. *Minerals* **2021**, *11*, 1334. [CrossRef]
14. Su, C.; Fan, D.; Jiang, J.; Sun, Z.; Liu, Y.; Song, W.; Wan, Y.; Yang, G.; Qiu, W. Self-Consistent Thermodynamic Parameters of Diopside at High Temperatures and High Pressures: Implications for the Adiabatic Geotherm of an Eclogitic Upper Mantle. *Minerals* **2021**, *11*, 1322. [CrossRef]

15. Tararushkin, E.V.; Pisarev, V.V.; Kalinichev, A.G. Equation of State, Compressibility, and Vibrational Properties of Brucite over Wide Pressure and Temperature Ranges: Atomistic Computer Simulations with the Modified ClayFF Classical Force Field. *Minerals* **2023**, *13*, 408. [CrossRef]
16. Liu, Z.; Wang, W.; Yin, K.; Hong, H.; Algeo, T.J.; Yin, Z.; Pan, Y.; Lu, Z.; Han, W.; Wang, Y.; et al. Gemological Characteristics of Lvwen Stone and Its Color Genesis. *Minerals* **2022**, *12*, 1584. [CrossRef]

minerals MDPI

Review

Some Remarks on the Electrical Conductivity of Hydrous Silicate Minerals in the Earth Crust, Upper Mantle and Subduction Zone at High Temperatures and High Pressures

Haiying Hu [1,*], Lidong Dai [1,*], Wenqing Sun [1], Yukai Zhuang [2], Kaixiang Liu [3], Linfei Yang [4], Chang Pu [5], Meiling Hong [1], Mengqi Wang [1], Ziming Hu [1], Chenxin Jing [1], Chuang Li [1], Chuanyu Yin [1] and Sivaprakash Paramasivam [6]

1 Key Laboratory of High-Temperature and High-Pressure Study of the Earth's Interior, Institute of Geochemistry, Chinese Academy of Sciences, Guiyang 550081, China; sunwenqing@vip.gyig.ac.cn (W.S.); hongmeiling@mail.gyig.ac.cn (M.H.); wangmengqi@mail.gyig.ac.cn (M.W.); huziming@mail.gyig.ac.cn (Z.H.); jingchenxin@mail.gyig.ac.cn (C.J.); lichuang@mail.gyig.ac.cn (C.L.); yinchuanyu@mail.gyig.ac.cn (C.Y.)
2 Institute of Atomic and Molecular Physics, Sichuan University, Chengdu 610065, China; zhuangyukai@scu.edu.cn
3 College of Material Science and Engineering, Guizhou Minzu University, Guiyang 550025, China; liukaixiang14@mails.ucas.ac.cn
4 Center for High Pressure Science and Technology Advanced Research, Beijing 100094, China; linfei.yang@hpstar.ac.cn
5 Department of Earth and Space Sciences, Southern University of Science and Technology, Shenzhen 518055, China; 11930803@mail.sustech.edu.cn
6 Centre for High Pressure Research, School of Physics, Bharathidasan University, Tiruchirappalli 620024, India; P.Sivaprakash@bdu.ac.in
* Correspondence: huhaiying@vip.gyig.ac.cn (H.H.); dailidong@vip.gyig.ac.cn (L.D.)

Citation: Hu, H.; Dai, L.; Sun, W.; Zhuang, Y.; Liu, K.; Yang, L.; Pu, C.; Hong, M.; Wang, M.; Hu, Z.; et al. Some Remarks on the Electrical Conductivity of Hydrous Silicate Minerals in the Earth Crust, Upper Mantle and Subduction Zone at High Temperatures and High Pressures. *Minerals* **2022**, *12*, 161. https://doi.org/10.3390/min12020161

Academic Editor: Julien Siebert

Received: 30 December 2021
Accepted: 26 January 2022
Published: 28 January 2022

Publisher's Note: MDPI stays neutral with regard to jurisdictional claims in published maps and institutional affiliations.

Abstract: As a dominant water carrier, hydrous silicate minerals and rocks are widespread throughout the representative regions of the mid-lower crust, upper mantle, and subduction zone of the deep Earth interior. Owing to the high sensitivity of electrical conductivity on the variation of water content, high-pressure laboratory-based electrical characterizations for hydrous silicate minerals and rocks have been paid more attention to by many researchers. With the improvement and development of experimental technique and measurement method for electrical conductivity, there are many related results to be reported on the electrical conductivity of hydrous silicate minerals and rocks at high-temperature and high-pressure conditions in the last several years. In this review paper, we concentrated on some recently reported electrical conductivity results for four typical hydrous silicate minerals (e.g., hydrous Ti-bearing olivine, epidote, amphibole, and kaolinite) investigated by the multi-anvil press and diamond anvil cell under conditions of high temperatures and pressures. Particularly, four potential influence factors including titanium-bearing content, dehydration effect, oxidation−dehydrogenation effect, and structural phase transition on the high-pressure electrical conductivity of these hydrous silicate minerals are deeply explored. Finally, some comprehensive remarks on the possible future research aspects are discussed in detail.

Keywords: electrical conductivity; impedance spectroscopy; water; titanium-bearing olivine; epidote; amphibole; kaolinite; high pressure

1. Introduction

As an important volatile matter, water is widely exists in the mid-lower crust, upper mantle, and subduction zone of the deep Earth interior, which is present as two main species in the water-bearing silicate minerals: (i) molecular water (H_2O) and structural hydroxyl (OH) [1]. For the existence of water as molecular water (H_2O), some typical minerals (e.g., chlorite, epidote, amphibole, kaolinite, et al.) are included, which are crucial to affect the water circulation in these regions of mid-lower crust, upper mantle, and subduction zones. As for some nominally anhydrous minerals (NAMs) (e.g., feldspar, quartz, olivine, clinopyroxene, orthopyroxene, garnet, wadsleyite, ringwoodite), most of them belong to a type of crucial silicate rock-forming minerals of the Earth's interior, in which the existence of water is as a form of the point defect of structural hydroxyl (OH) in the hydrous minerals. As usual, it is well known that the hydrous minerals and rocks are the major carrier of water in the subducted geotectonic environment and NAMs can carry trace amounts of water into the deep uppermost mantle. Since the first theoretical calculations of electrical conductivity on hydrous olivine from Karato (1990) based on the proton diffusion of Nernst–Einstein equation, there are a large quantity of electrical conductivity results to be reported from the laboratory-based electrical conductivity experiments, theoretical calculations of molecular dynamics, and filed geophysical observation of magnetotellurics (or electromagnetic depth soundings) [2–8]. Furthermore, by virtue of these acquired high-pressure laboratory-based conductivity results, we can extrapolate the water content, distribution state, storage form, and migration mechanism of water in the deep Earth interior.

Recently, some available research found that the presence of water, either the trace molecular water (H_2O) or the structural hydroxyl (OH) in hydrous mineral, plays a crucial role in many pressure-dependent physicochemical properties and their corresponding transport processes of deep Earth interior, such as electrical conductivity [9,10], elastic wave dispersion and its attenuation [11,12], grain–growth kinetics [13,14], elemental diffusion coefficient [15,16], dislocation creep [17,18], and dynamic recrystallization [19,20]. Whereas, the electrical conductivity of hydrous silicate minerals and rocks are highly sensitive to temperature, pressure, oxygen fugacity, crystallographic anisotropy, water-bearing content, dehydration effect, iron content, trace element of titanium-bearing content, oxidation–dehydrogenation effect, structural phase transition, etc. As pointed out by Karato (1990) [2], Huang et al. (2005a, b; 2017) [21–23], Wang et al. (2006) [24], Dai and Karato (2009a, b, c; 2014a, b, c, d; 2020) [25–31], Karato and Dai (2009) [32], Hu et al. (2017, 2018) [33,34], Dai et al. (2020) [35], and He et al. (2021) [36], water content is one of most important influential factor on the electrical properties of minerals and rocks at high-temperature and high-pressure conditions. Water can enhance several orders of magnitude in the electrical conductivity of hydrous minerals and rocks, and its effect is substantial.

With the improvement and development of high-pressure experimental techniques and measurement methods for electrical conductivity, more and more conductivity results were obtained in order to systematically investigate the mineralogical composition, the electrical conductivity—depth profile, and discontinuity of cycle-layered structure of Earth interior in the globally and regionally geotectonic units. In the last several decades, the electrical conductivity of hydrous minerals and rocks have been performed by virtue of electrochemical alternating current (*AC*) impedance spectroscopy using all of these available high-pressure pieces of equipment, including autoclave, piston-cylinder, multi-anvil press, and diamond anvil cell. In this work, we focused some of the most recent developments in electrical conductivity of hydrous minerals in the multi-anvil press and diamond anvil cell. In this review paper, some potential influence ingredients from the titanium-bearing content, dehydration effect, oxidation−dehydrogenation effect, and structural phase transition on the electrical conductivity of hydrous minerals are discussed in detail.

2. High-Pressure Apparatus for the EC Measurements and FT-IR Observation

In the last several decades, in order to realize the electrical conductivity measurements of hydrous minerals and rocks at HP-HT conditions, many researchers have already designed and developed their high-pressure experimental platforms, such as autoclave, piston-cylinder, multi-anvil press, and diamond anvil cell.

Recently, Shan et al. (2021) from the Key Laboratory of High-Temperature and High-Pressure Study of the Earth's Interior (HTHPSEI), Institute of Geochemistry, Chinese Academy of Sciences, successfully designed and developed one new high-pressure experimental platform for the measurement of electrical conductivity for water-bearing porous rocks using autoclave, which can be widely applied to a shallow Earth crustal environments under conditions of high temperature, high pressure, and controlling the liquid water saturation [37]. In order to test its accuracy and precision, the electrical conductivity of water-bearing sandstone were performed under controlled temperature ranges from 303 to 423 K and the fixed saturation degree ranges from 36 to 100% in the liquid water at 2.0 MPa. Shan et al. (2021) found that the electrical characterization technique under the relatively lower-temperature and lower-pressure conditions is widely applicable to a series of high porosity-containing hydrous rocks in the geothermal/oil research.

Early in 2009, Dai and his collaborators have already set up one high-pressure platform of high-pressure electrical conductivity measurement in order to check the influence of oxygen fugacity on the grain boundary electrical conductivity of rocks using YJ-3000t multi-anvil high-pressure apparatus in HTHPSEI, at the Institute of Geochemistry, Chinese Academy of Sciences [38]. In their designed experimental method, as an example of dry peridotite, the grain interior, grain boundary, and total electrical conductivities for dry synthetic peridotite were performed by virtue of the electrochemical *AC* impedance spectroscopy technique at high temperature, high pressure, and controlled oxygen fugacities. To efficiently adjust and control the oxygen fugacity during the process of electrical conductivity measurements in the sample chamber, five typical solid buffers were employed, i.e., Fe_3O_4 + Fe_2O_3, Ni + NiO, Fe + Fe_3O_4, Fe + FeO, and Mo + MoO_2. With the rise of oxygen fugacity, the grain boundary electrical conductivity of dry peridotite will increase, accordingly. Further, the functional relationship between the grain boundary electrical conductivity of dry peridotite and oxygen fugacity was successfully established under conditions of fixed pressure of 2.0 GPa and temperature of 1173 K. One detailed experimental measurement platform was displayed in Figure 1, which is adopted to measure the electrical conductivity of minerals and rocks in HTHPSEI, conducted at the Institute of Geochemistry, Chinese Academy of Sciences, in HP-HT conditions. For this representative conductivity measurement platform, there are three dominant pieces of equipment to be used. The high-temperature and high-pressure environments were realized by the YJ-3000t multi-anvil press, as shown in Figure 1a. Figure 1b stands for our Solartron-1260 impedance/gain-phase analyzer, which is also in conjunction with Solartron-1296 impedance spectroscopy analyzer interface in order to enhance the measurement range of sample resistance up to 10^{12} Ohm. As usual, the variation of water content is checked for the initial and recovered samples before and after each electrical conductivity experiments, and thus, the Vertex-70V vacuum high-resolution and high-precision Fourier-transform infrared spectroscopy (FT-IR) analyzer will be employed to efficiently determine the trace structural water in minerals and rocks, as displayed in Figure 1c.

It is well known that many high-pressure mineral physiochemical properties for minerals and rocks are extremely sensitive to the variation of trace structural water at HP-HT conditions. To precisely check the water content in the sample, the high-resolution and high-precision FT-IR measurement is indispensable to outline it. During each electrical conductivity measurement, the FT-IR analysis is performed within the wide wavenumber ranging from 600 to 8000 cm^{-1}. In this FT-IR measurement platform, it consists of two corresponding pieces of the vacuum Vertex-70V spectrometer and the high-resolution Hyperion-1000 infrared microscope. For each representative FT-IR measurement, the sample must be cut thinner and polished enough so as to assure enough of the transmitted light across the experimental sample. Thus, the thickness in the sample was determined according to the different transparency of each corresponding mineral, whose magnitude is also determined by the high-precision Olympus microscope. The water content of sample was acquired using the unpolarized radiation of mid-IR light source with the IR absorption mode. Then, the square of IR aperture is adjusted to the magnitude of 100×100 μm^2, which is equipped with a high-performance infrared detector composed of a MCT (Mercury-Cadmium-Telluride) pseudobinary alloy semiconducting material. In the stage of the Hyperion-1000 infrared microscope, one circular slice of white and transparent calcium fluoride (CaF_2) or potassium bromide (KBr) with its diameter of 5 cm and thickness of 1 mm was installed as the beam splitter in our IR measurement. In order to efficiently avoid the inevitable effect from the heterogeneity of water distribution in the sample, at least five sets of FT-IR spectra were gathered from different selected positions in each corresponding mineral surface, and then the water content is precisely determined according to the averaging value. For a selected total absorbance of OH groups between wavenumber ranges of 3000–3800 cm^{-1}, each FT-IR spectroscopy of mineral was conducted at a predetermined 64, 128, 256, and 512 scans accumulated for each sample surface. Except for high-pressure electrical conductivity measurements, the ultrasonic elastic wave velocity and synchrotron X-ray diffraction of hydrous mineral are also determined the water content before and after high-temperature and high-pressure experiments, as illustrated in detail by Gwanmesia et al. (2020) for hydrous wadsleyite using our FT-IR experimental platform [39].

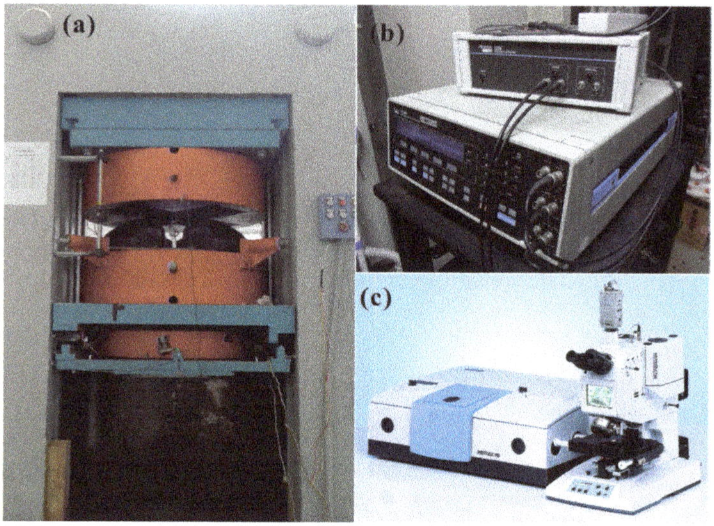

Figure 1. High-pressure experimental platform for the measurements of electrical properties of minerals and rocks using the electrochemical *AC* impedance spectroscopy in HTHPSEI, Institute of Geochemistry, Chinese Academy of Sciences. In here, (**a**) stands for the YJ-3000t multi-anvil apparatus, (**b**) stands for the combined interface of Solartron-1260 and Solatron-1296 Impedance/Gain-phase analyzers, and (**c**) stands for the Vertex-70V vacuum high-resolution and high-precision FT-IR analyzer, respectively.

Figure 2 shows the representative FT-IR spectra for hot-pressed sintering polycrystalline olivine aggregates within the wavenumber range from 3000 to 3800 cm^{-1}. According to the detailed calculating method of water content for hydrous olivine single crystals and hot-pressed sintering polycrystalline aggregates from Dai and Karato (2014a, b, c, d), some hydrogen-related defects are displayed at the wavenumber ranges from 3000 to 3750 cm^{-1} [28–31]. Therefore, the integration of FT-IR spectroscopy is conducted at a given wavenumber range, and the water content in the hot-pressed sintering polycrystalline olivine aggregates was determined using an equation proposed by Paterson (1982) [40],

$$C_{\mathrm{OH}} = \frac{B_{\mathrm{i}}}{150\xi} \int \frac{K(\nu)}{(3780-\nu)} d\nu \tag{1}$$

where, C_{OH} is the molar concentration of the OH group (its correspondent unit is the ppm wt% or $H/10^6$ Si of water), B_{i} is the density factor (4.39×10^4 cm $H/10^6$ Si), ξ is the orientation factor (In case of the hot-pressed sintering polycrystalline aggregates or rock sample, the value is assigned as 1/3; Whereas, in case of mineral single crystal sample, the value is assigned as 1/2), and $K(\nu)$ is the absorption coefficient in cm^{-1} at the wavenumber ν (cm^{-1}). The integration was conducted over the wavenumber ranges from 3100 to 3750 cm^{-1}. The measurement error of water content is less than 10%, which is mainly originated from the uncertainty of sample thickness measurement and the inevitable heterogeneity of each averaging FTIR spectroscopy of a standard water content calibration for hydrous hot-pressed sintering polycrystalline olivine aggregates.

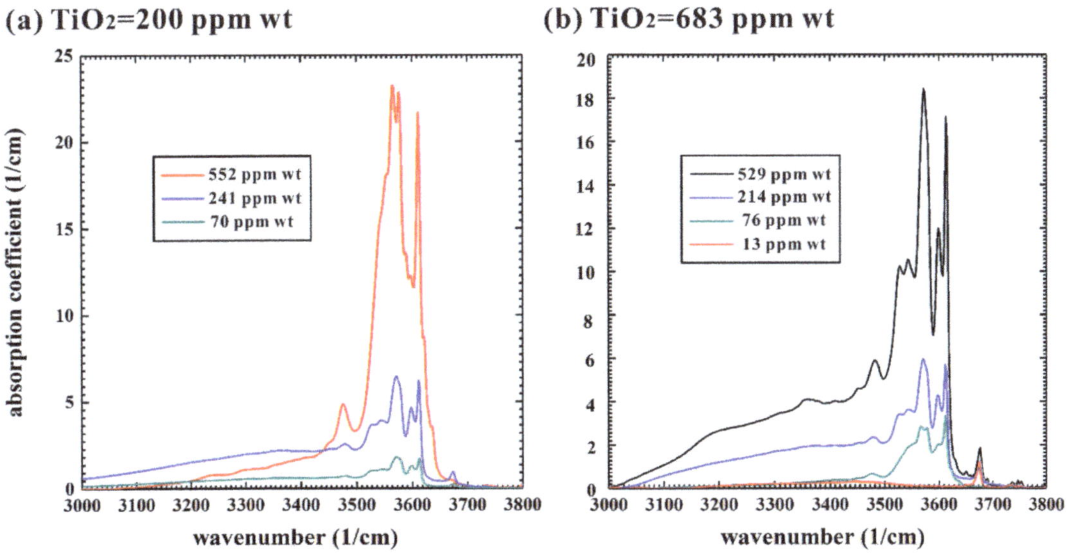

Figure 2. Representative FTIR spectra for hot-pressed sintering polycrystalline olivine aggregates within the wavenumber range from 3000 to 3800 cm^{-1}. Here, (**a**) displays a series of Ti-poor sample with three different water contents (552, 241, and 70 ppm wt); (**b**) displays a series of Ti-rich sample with four different water contents (529, 214, 76, and 13 ppm wt). The titanium content in polycrystalline olivine aggregates (200 and 683 ppm wt) was precisely checked by the inductively coupled plasma mass spectrometry (ICP-MS) analysis in the Department of Earth and Planetary Sciences, Yale University (reproduced with permission from the reference of Dai and Karato [9]; published by the American Geophysical Union, 2020).

In the last several years, many electrical conductivity results of hydrous minerals and rocks (e.g., Fe-bearing amphibole, epidote, garnet, gabbro, basalt, mudstone, phyllite, et al.) in the mid-lower Earth crust and subduction zone have already investigated on this YJ-3000t multi-anvil high-pressure apparatus in conjunction with the electrochemical *AC* impedance spectroscopy and FT-IR spectroscopy measurements at high temperatures and high pressures [34,35,41–45]. The influences of dehydration effect and dehydrogenation effect on hydrous minerals and rocks were explored in detail, for which, all of them were widely applied to disclose the high conductivity anomalies in the regions of subduction zone and stable continental crust. Except for all of these acquired electrical conductivities of hydrous minerals and rocks, a series of electrical conductivities of dry geological specimens from the Earth's crust and upper mantle were paid much attention at high temperature and high pressure, such as albite, K-feldspar, alkali feldspar solid solutions, anorthite, olivine, orthopyroxene, diopside, almandine-rich garnet, peridotite, lherzolite, pyroxenite, pyroxene andesite, quartz andesite, granite, carbonaceous slate, schist, granulite, gneiss, eclogite, etc., in this multi-anvil high-pressure apparatus [46–71]. All of these influential factors (e.g., frequency, temperature, oxygen partial pressure, pressure, water-bearing content, iron-bearing content, anisotropy of crystallographic axis, segregation effect of grain boundary, content of alkali metallic ion, salinity-bearing, water-bearing fluids, etc.) have systematically been investigated on the basis of this *in-situ* multi-anvil high-pressure apparatus.

In addition, in order to realize the electrical conductivity measurements at much higher pressure conditions, we successfully designed and constructed a new experimental platform of electrical conductivity using diamond anvil cell in HTHPSEI, at the Institute of Geochemistry, Chinese Academy of Sciences. A detailed experimental setup and its corresponding cross-sectional assembly diagram of diamond anvil cell for high-pressure electrical conductivity measurements, as displayed in detail in Figure 3. A detailed experimental method and measurement procedure is presented in Dai et al. (2019) [72]. Furthermore, we well combined this *in-situ* high-pressure EC measurement system with other crucial research means, including high-pressure Raman spectroscopy, high-resolution transmission electron microscopy, atomic force microscopy, and first-principles theoretical calculations in order to check high-pressure electrical transport, vibrational, and structural properties. Some representative pressure mediums including the mixture of volume ratio (4:1) for the methanol and ethanol, liquid argon, liquid helium, etc., were selected to realize different hydrostatic environments in the sample chamber of diamond anvil cell. To note, more detailed high-pressure research was deeply explored during the process of both compression and decompression. By virtue of these electrical, spectroscopic, and theoretically calculating studies, a series of physiochemical behaviors of the pressure-induced structural phase transformation and dehydration effect on the hydrous minerals (e.g., kaolinite, epsomite, goethite, chalcanthite, gypsum, etc.) have already studied from Dai Lidong's high-pressure research group in HTHPSEI, at the Institute of Geochemistry, Chinese Academy of Sciences [73–77]. In addition, some important semiconducting materials including the binary metallic oxides, AB-type structural layered metallic compounds, AB_2-type structural transition-metal dichalcogenides (TMDs), and ABO_3-type structural perovskite compounds are paid attention in regards to their structural phase transition, amorphization, and metallization, and investigated under conditions of HP-HT conditions in detail [78–97].

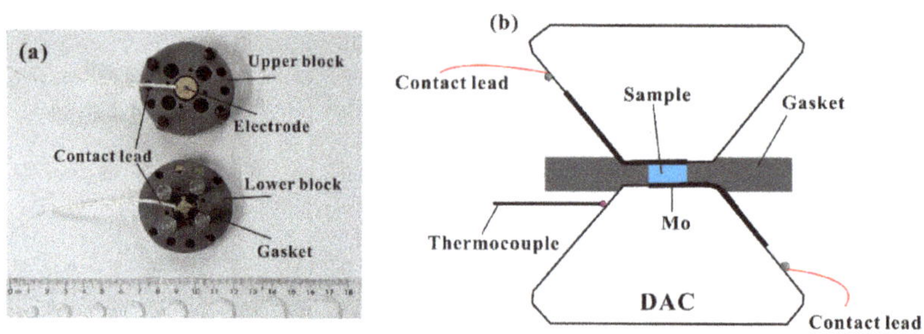

Figure 3. Measurement assemblage of sample for electrical conductivity measurements in the diamond anvil cell (DAC) at high pressure. In here, (**a**) stands for a detailed configuration for two symmetric plate electrodes integrated between upper and lower blocks of diamond anvils; (**b**) stands for the cross-sectional diagram of the DAC used for high-pressure impedance spectroscopy measurements.

3. Experimental Theory and Measurement Methods

It is well known that the electrochemical *AC* impedance spectroscopy method is one of most efficient technique to measure the electrical conductivity of hydrous minerals and rocks at conditions of high temperatures and high pressures [98–106]. Before we launched a complex impedance spectroscopy measurement, the *AC* signal voltage ranges from 10 mV to 3 V and the scanning frequency ranges from 10^{-5} Hz to 3.2×10^7 Hz need to be predesignated by virtue of Z-Plot program in the complex impedance spectroscopy analyzer. The measurement mode of controlling signal voltage and scanning frequency was selected to obtain the complex impedance spectroscopy of sample. As usual, for a typical single-crystal mineral, the acquired complex impedance spectroscopy consisted of one semi-circular arc of grain interior, as well as the electric transport process of sample—electrode interface polarization. However, for a special polycrystalline aggregates or rock, the acquired complex impedance spectroscopy consists of one semi-circular arc of grain interior, one semi-circular arc of grain boundary, as well as the electric transport process of sample—electrode interface polarization. Accordingly, by fitting all of these acquired complex impedance spectra using the appropriate equivalent electric circuit, the grain interior electrical conductivity, grain boundary electrical conductivity and the sample—electrode interface polarization resistance will be successfully obtained in turns. For each individual complex impedance spectroscopy, four characteristic parameters including the real part (Z′), the imaginary part (Z″), magnitude (|Z|), and phase angel (θ) will be obtained at the same time. The relationship between the real part, the imaginary part, magnitude, and phase angel can be expressed as,

$$\theta = \left|\tan^{-1}(Z''/Z')\right. \quad (2)$$

$$|Z| = \left[(Z')^2 + (Z'')^2\right]^{1/2} \quad (3)$$

Representative complex impedance spectra for natural hydrous Fe-bearing amphibole single crystals from Sichuan province of southwest China at conditions of 1.0 GPa, 623–1173 K, and frequency range of 10^{-1}–10^6 Hz before and after the occurrence of the oxidation–dehydrogenation reactions, which are illustrated in Figure 4 in detail.

More detailed descriptions on experimental theory and measurement methods for the impedance spectroscopy are presented in our previous review paper [35]. The equivalent electric circuit was selected to fit the impedance spectroscopy of sample, which is composed of some fundamental electronic elements (e.g., resistor, capacitor, inductor, constant phase element (CPE), gerischer element, Warburg element, etc.) [107–109]. After that, the electrical conductivity of sample was obtained, and the calculating formula is expressed as,

$$\sigma = \frac{1}{\rho} = \frac{L}{(R \times S)} \quad (4)$$

Here, σ stands for the electrical conductivity (S/m), ρ stands for the electrical resistivity (m/S), L stands for the sample height (m), and S stands for the cross-sectional area of sample (m^2). As usual, the relationship between the electrical conductivity of hydrous mineral and temperature satisfied with an Arrhenius relation, i.e., with the rise of temperature, and the electrical conductivity of sample will be increased, accordingly. As noted from a modified Arrhenius relation reported by Dai and Karato (2020) [9], the influences of water content and oxygen fugacity on the electrical conductivity of dry and hydrous Ti-doped polycrystalline olivine aggregates can be expressed as,

$$\sigma = \sigma_{0(\mathrm{Dry})}\left(\frac{f_{O_2}}{f_{O_2 0}}\right)^{q_{\mathrm{Dry}}} \exp\left(-\frac{\Delta H_{\mathrm{Dry}}}{kT}\right) + \sigma_{0(\mathrm{Hydrous})}\left(\frac{C_W}{C_{W0}}\right)^{r}\left(\frac{f_{O_2}}{f_{O_2 0}}\right)^{q_{\mathrm{Hydrous}}} \exp\left(-\frac{\Delta H_{\mathrm{Hydrous}}}{kT}\right) \quad (5)$$

where $\sigma_{0(\mathrm{Dry})}$ and $\sigma_{0(\mathrm{Dry})}$ stand for the pre-exponential factor of dry and hydrous Ti-bearing olivine aggregates (S/m), respectively; ΔH_{Dry} and $\Delta H_{\mathrm{Hydrous}}$ stands for the activation enthalpy of dry and hydrous titanium-bearing olivine aggregates (eV), respectively; f_{O_2} stands for the oxygen fugacity as a standardized oxygen buffer ($f_{O_2 0}$); q_{Dry} and q_{Hydrous} stands for the exponential factor characterizing the dependence of oxygen fugacity on the electrical conductivity of dry and hydrous Ti-bearing olivine aggregates, respectively; C_w stands for the water content as the standardized reference water-bearing content (C_{w0}); r stands for the exponential factor of water content; T stands for the absolute temperature (Kelvin); and k stands for the Boltzmann constant.

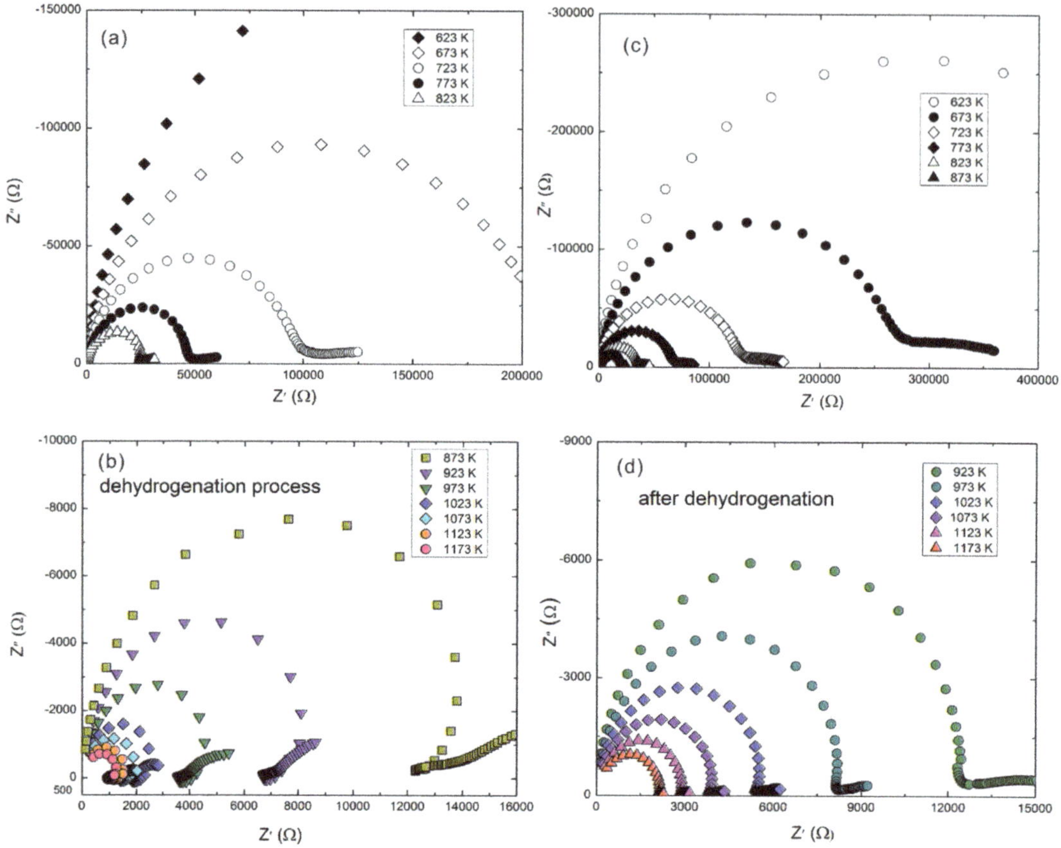

Figure 4. Representative complex impedance spectra for natural hydrous Fe-bearing amphibole single crystals from Sichuan province of southwest China at conditions of 1.0 GPa, 623–1173 K and the frequency ranges from 10^{-1} to 10^6 Hz. Here, (**a**,**b**) represent complex impedance spectra during the 1st heating cycle before the dehydrogenation process, respectively; (**c**,**d**) represent complex impedance spectra during the equivalent state of 3rd cooling cycle after the oxidation−dehydrogenation reaction, respectively. It is clear that the variation of impedance semi-circular arc from the inductive loop to normally "standardized" small tail within the low frequency regime of ~10^{-1}−10^3 Hz at the critical temperature of ~873 K reveals the occurrence of an oxidation−dehydrogenation reaction. The equivalent circuit composed of two individually parallel resistance and constant phase element in the series connection was chosen to fit these impedance arcs for (**a**,**c**,**d**), respectively; the equivalent circuit composed of two individually parallel resistance and inductive reactance in the series connection was chosen to fit these impedance arcs for (**b**) (reproduced with permission from the reference of Hu et al. [34]; published by Elsevier, 2018).

4. Electrical Conductivity of Hydrous Minerals

In this section, recently reported results of the EC of Ti-bearing hydrous olivine aggregates from the KIWI 1000-ton Kawai-type multi-anvil apparatus in the Karato High-pressure Laboratory, Department of Earth and Planetary Sciences, Yale University, by Dai and Karato (2020) were chosen as examples to investigate the effect of structural hydroxyl (OH) and transition metallic element of titanium on the electrical conductivity hydrous minerals at high temperatures and high pressures [9]. By virtue of YJ-3000t multi-anvil press and Solartron-1260 impedance analyzer in HTHPSEI, Institute of Geochemistry, Chinese Academy of Sciences, the EC measurements of epidote and amphibole by Hu et al.

(2017; 2018) were performed in order to study the influences of dehydration effect and oxidation–dehydrogenation effect from the molecular structural water (H_2O) on the electrical conductivity of hydrous minerals at HP-HT conditions [33,34]. In the end, Hong et al. (2022) conducted the electrical conductivity of hydrous kaolinite in conjunction with *in-situ* high-temperature and high-pressure electrochemical *AC* impedance spectroscopy and Raman scattering results using the diamond anvil cell in HTHPSEI, Institute of Geochemistry, Chinese Academy of Sciences, and the pressure-induced structural phase transition for hydrous minerals was explored in detail [73].

4.1. Electrical Conductivity of Ti-Bearing Hydrous Olivine

As major rock-forming NAMs, silicate mineral of olivine's volume proportion occupies approximately up to 60% in a representative upper-mantle mineralogical composition. Thus, a large number of previously available reported electrical transport characterization—depth profiles were successfully established on the base of the laboratory-based electrical conductivity of olivine results at conditions of high temperature and high pressure. In recent decades, all of those potential influence factors including frequency, temperature, oxygen partial pressure, pressure, segregation effect of grain boundary, interconnected high conductive impurity phases (e.g., magnetite, chromite, etc.), salinity-bearing fluid, Ti-bearing content, water-bearing content, iron-bearing content, anisotropy of crystallographic axis, etc. have been paid more attention in regards to natural preferred olivine single crystal, polycrystalline olivine aggregates, and hot-pressed sintering synthetic olivine aggregates. According to these acquired electrical conductivity results on the EC of hydrous olivine, the cause of high conductivity anomaly in some typical regions of asthenosphere of Earth mantle and subduction zone was explored in detail.

Hydrogen as a form of point defect dissolved in the crystalline structure of olivine can enhance several magnitude of orders in the electrical conductivity by virtue of the diffusion of some hydrogen-related defects [30]. Due to the complexity of the concentration and its corresponding mobility in the electric charge carrying species for these hydrogen-related defects, the electrical conductivity of hydrous single crystals along with [100], [010], and [001] of three main crystallographic orientations exists as an obvious anisotropy when the experimental temperature is higher than 1000 K at the given pressure of 4.0 GPa [28]. However, recent research results on the redox-influenced elastic wave velocity of upper-mantle mineral reported by Cline et al. (2018) find that some anionic impurities of transition metallic elements including the charged tetravalent titanium (Ti^{4+}) and trivalent chromium (Cr^{3+}) dissolved in the hydrous olivine are possibly responsible for low-velocity or high-attenuation behaviors in the upper mantle under conditions of the confining pressure of 200 MPa and temperatures up to 1473 K [110]. As pointed out from the experimental results of water solubility at 3.0 GPa and 1323 K reported by Tollan et al. (2017), the defect reaction of dissolved titanium as a rutile form, and hydrogen in olivine can be expressed as [111],

$$TiO_2 + H_2O + \text{Olivine} = \left[Ti_M^{\bullet\bullet} + (2H)_{Si}''\right]_{\text{Olivine}} + Mg_2Si_2O_6 \tag{6}$$

where the mark signal of $\left[Ti_M^{\bullet\bullet} + (2H)_{Si}''\right]_{\text{Olivine}}$ stands for two titanium electrons on the lattice of metallic site in olivine and two hydrogen vacancies on the lattice of silicon site in olivine. If this defect reaction occurs, the interaction between titanium and hydrogen becomes stronger, and the mobility and concentration of hydrogen in olivine will be reduced, accordingly. Thus, Dai and Karato (2020) firstly synthesized the high-purity polycrystalline olivine aggregates with two different TiO_2-bearing contents of 200 and 683 ppm wt using the Ar + H_2/CO_2 mixed gas controlling oxygen fugacity at the high-temperature furnace [9]. Then, the mixtures of brucite and talc with its correspondent weight ratio of 1:6 as the water source were employed to perform a hot-pressed sintering in the KIWI 1000-ton Kawai-type multi-anvil apparatus at conditions of 4.0 GPa and 1373 K for 3 hrs. A series of initial polycrystalline olivine aggregates with different water-bearing and titanium-bearing contents were successfully synthesized. Electrical conductivities of water-poor and water-rich polycrystalline olivine aggregates with two different TiO_2-

bearing contents of 200 ppm wt and 683 ppm wt were performed under conditions of 4 GPa, 873–1273 K, and the Ni–NiO oxygen buffer, respectively. The influence of water contents on the electrical conductivity of polycrystalline olivine aggregates with its correspondent TiO_2-bearing content of 200 ppm wt are shown in Figure 5a,b, and the influence of water contents on the electrical conductivity of polycrystalline olivine aggregates with its correspondent TiO_2-bearing content of 683 ppm wt were displayed in Figure 5c,d, respectively.

All of these obtained electrical conductivity result dependencies on the water content and oxygen fugacity for Ti-doped polycrystalline olivine aggregates were fitted by virtue of Equation (5). The exponential factors between the electrical conductivity along with the variation of water content for Ti-poor and Ti-rich samples (r) were determined as 0.51 $\pm$ 0.18 and 0.98 $\pm$ 0.21, respectively. In comprehensive considerations with previously available electrical conductivity results with different water contents from the natural San Carlos olivine reported by Wang et al. (2006) [24], as well as hot-pressed sintering titanium-free olivine aggregates reported by Dai and Karato (2014d) [31], the effect of titanium content on the electrical conductivity of dry polycrystalline olivine aggregates is very larger, and whereas, the effect of titanium content on the electrical conductivity of hydrous polycrystalline olivine aggregates becomes extremely feeble.

On the other hand, Dai and Karato (2020) also performed the EC measurements for the anhydrous and hydrous polycrystalline olivine aggregates with a fixed TiO_2 content of 683 ppm wt under conditions of 4.0 GPa, 873–1273 K and different oxygen fugacities, as illustrated in detail in Figure 6 [9]. The water contents in the anhydrous and hydrous olivines are 13 and 458 ppm wt, respectively. Three representative solid oxygen buffers including Re–ReO_2, Ni–NiO, and Mo–MoO_2 were selected in order to control oxygen fugacity in the sample chamber during the process of electrical conductivity measurements. Completely different dependence relations of water-free and water-rich electrical conductivity in the Ti-rich polycrystalline aggregate olivines on the variation of oxygen fugacity were observed. In the case of anhydrous condition, the electrical conductivity of sample increased with the rise of oxygen fugacity. On the contrary, the electrical conductivity of the sample reduced with the rise of oxygen fugacity under hydrous condition. Furthermore, the average exponential factor values for the electrical conductivity of anhydrous and hydrous polycrystalline olivine aggregates along with the variation of oxygen fugacity (q) are determined as 0.097 $\pm$ 0.008 and -0.089 $\pm$ 0.006, respectively.

In addition, in order to reasonably explain all of the above-mentioned high-pressure electrical conductivity results, it is the first time for some titanium-related defects (such as $Ti_M^{\bullet}$, $Ti_M^{\bullet\bullet}$, etc. Here, $Ti_M^{\bullet}$ denotes the trivalent titanium ion of metallic site in the olivine crystalline lattice, $Ti_M^{\bullet\bullet}$ denotes the tetravalent titanium ion of metallic site in the olivine crystalline lattice) and hydrogen-related defects (such as $(3H)'_{Si}$, $(4H)''_{Si}$, etc. Here, $(3H)'_{Si}$ denotes three hydrogen vacancies of silicon site in the olivine crystalline lattice, $(4H)''_{Si}$ denotes four hydrogen vacancies of silicon site in the olivine crystalline lattice) in Ti-doped and hydrogen-doped olivine aggregates, as put forward by Dai and Karato (2020) [9]. All of these available positive and negative dependence relations of oxygen fugacity on the electrical conductivity Ti-bearing olivine aggregates with different water contents are resulted in the distinction of electrical transport conduction, which is in consistent with the previously reported electrical conductivity results for San Carlos olivine, orthopyroxene, diopside, pyrope-rich garnet, and almandine-rich garnet in the upper mantle, as well as for wadsleyite in the mantle transition zone [25,27,41,51,58,59].

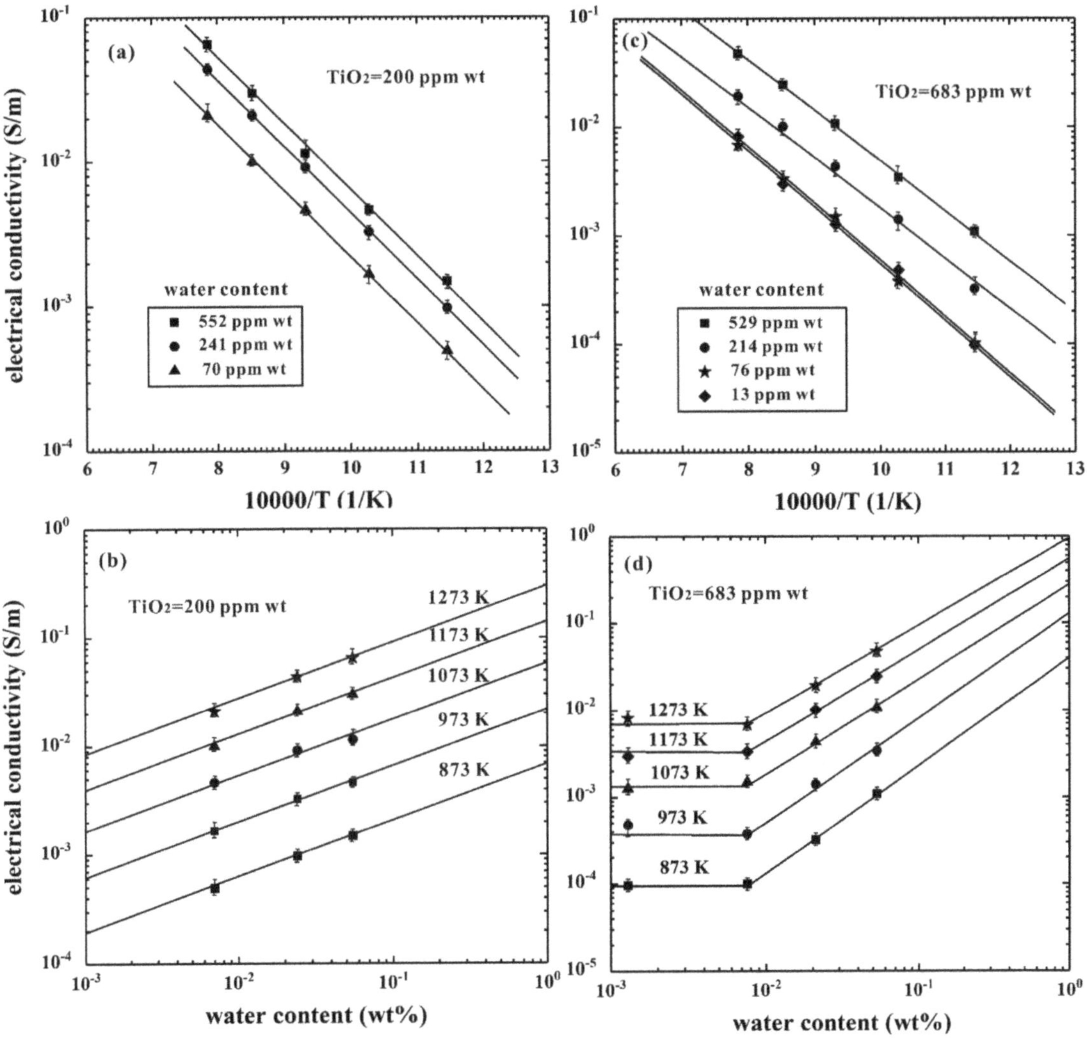

Figure 5. Electrical conductivity of hot-pressed polycrystalline olivine aggregates with different TiO_2 contents under conditions of 4 GPa, 873–1273 K, and the Ni–NiO oxygen buffer. Here, (**a**,**c**) stand for the dependence of electrical conductivity on the inverse temperature for olivines with its correspondent TiO_2 contents of 200 and 683 ppm wt, respectively; (**b**,**d**) stand for the dependence of electrical conductivity on the water content for olivine samples with its correspondent TiO_2 contents of 200 and 683 ppm wt, respectively. The exponential factors between the electrical conductivity for Ti-poor and Ti-rich olivines along with the variation of water content were determined as 0.51 ± 0.18 and 0.98 ± 0.21, respectively (reproduced with permission from the reference of Dai and Karato [9]; published by the American Geophysical Union, 2020).

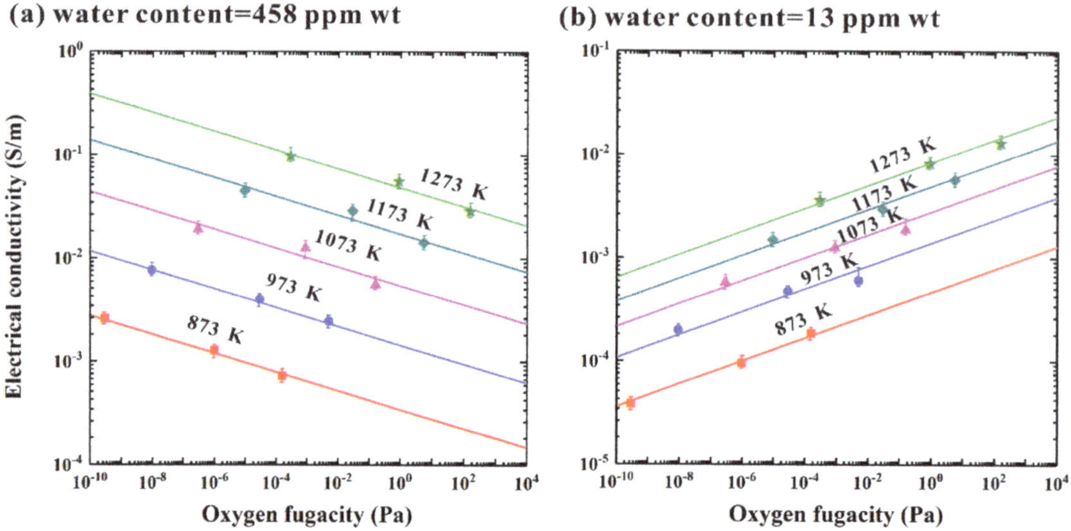

Figure 6. The electrical conductivity of dry (13 ppm wt) and hydrous (458 ppm wt) polycrystalline olivine aggregates with the correspondent TiO_2 content of 683 ppm wt as a function of the variation of oxygen fugacity at 4.0 GPa, temperatures range from 873 to 1273 K and controlled different oxygen fugacities. In here, (**a**) stands for the high Ti-doped hydrous sample, and as well as (**b**) stands for the high Ti-doped dry sample, respectively. The oxygen fugacity is efficiently controlled by three solid oxygen buffers including the rehnium and rehnium dioxide (Re–ReO_2), nickel and nickel oxide (Ni–NiO), and molybdenum and molybdenum dioxide (Mo–MoO_2). The FTIR measurement results indicated that the water contents of dry and hydrous polycrystalline olivine aggregates are 13 and 458 ppm wt, respectively (reproduced with permission from the reference of Dai and Karato [9]; published by the American Geophysical Union, 2020).

4.2. Electrical Conductivity of Hydrous Epidote

As a typical hydrous rock-forming mineral, the episode belongs to a nesosilicate structural epidote family with the Ca−rich and Al−rich complex chemical compositions, whose corresponding chemical molecular formula can be expressed as $Ca_2Al_{3-x}Fe_xSi_3O_{13}H$ (The value of x is fell in between the range from 0 to 1.0). As usual, hydrous epidote is of the dominant reaction product during the process of the phase transformation between greenschist to blueschist facies among the regional metamorphism process. At the same time, the abundance of epidote will be progressively increased with the rise of pressure, and its volume percentage of highest abundance can reach up to 35%, occurring in epidote-blueschist facies. At a depth higher than 80 km in the region of subducting oceanic crust, the hydrous epidote is still stabilized, which was considered as an important water reservoir to be responsible for the fluid flux within the depth ranges of 100−120 km of the deep mantle wedge region [112]. Thus, the released water-bearing fluid from the dehydration effect of epidote is crucial to reasonably explain the high conductivity anomaly in the region of the deep mantle wedge.

Hu et al. (2017) measured the electrical conductivity of hydrous epidote at the pressure ranges from 0.5 to 1.5 GPa and temperatures ranges from 573 to 1273 K, as displayed in detail in Figure 7 [33]. A relatively feeble pressure effect on the electrical conductivity of hydrous epidote was observed in comparison with the experimental temperature. The occurrence on the dehydration reaction of sample was well confirmed by virtue of the moderate discontinuity at the critical temperature point of ~1073 K. The dehydration of hydrous epidote results in the breakdown of sample and the increase of electrical conductivity up to 1 S/m when the experimental temperature was enhanced to 1273 K.

Such a high electrical conductivity value (1 S/m) is owing to the released water-bearing fluid during the process of initial dehydration reaction of epidote. After the dehydration of hydrous epidote was thoroughly completed, a remarkable reduction in the electrical conductivity of the sample is highly related to the appearance of newly coexisting multiphase mineralogical assemblage (grossular garnet, anorthite, quartz, and hematite) with a high electrical resistivity, as well as the escape of water-bearing fluid from the dehydration product of residual epidote.

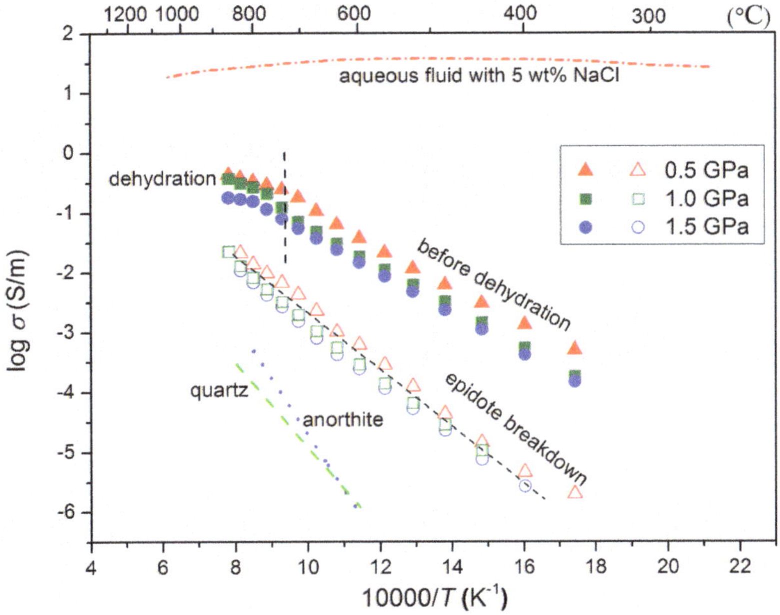

Figure 7. Pressure dependent electrical conductivity of hydrous epidote at the pressure ranges from 0.5 to 1.5 GPa and temperatures ranges from 573 to 1273 K. The dehydration reaction of hydrous epidote occurred at the critical temperature point of ~1073 K. The electrical conductivity of hydrous epidote is well reproducible before and after the occurrence of dehydration reaction at given pressure ranges from 0.5 to 1.5 GPa. Previously, acquired electrical conductivity results on the hot-pressed sintering polycrystalline quartz from Bagdassarov and Delépine (2004) [113], the synthetic polycrystalline anorthite from Hu et al. (2015) [60], as well as the pure aqueous fluid with the 5% of weight percentage of sodium chloride from Sinmyo and Keppler (2017) [114] are also compared in detail (reproduced with permission from the reference of Hu et al. [33]; published by the American Geophysical Union, 2017).

At the same time, Hu et al. (2017) also found that the electrical conductivity of hydrous epidote is well reproducible before and after the occurrence of dehydration reaction at given pressure ranges from 0.5 to 1.5 GPa [33]. In comprehensive comparisons with previously acquired electrical conductivity results on the hot-pressed sintering polycrystalline quartz reported by Bagdassarov and Delépine (2004) [113], synthetic polycrystalline anorthite reported by Hu et al. (2015) [60], as well as the pure aqueous fluid of sodium chloride with the 5% of weight percentage reported by Sinmyo and Keppler (2017) [114], one new origin for the explanation of high conductivity anomaly at the depth ranges from 70 to 120 km was put forward on the base of the released water-bearing fluid during the process of epidote dehydration at high temperature and high pressure. Furthermore, some recently reported electrical conductivity and elastic wave velocity results on hydrous halogen-bearing amphibole, glaucophane, Liebermannite, chlorite, and lawsonite from Manthilake et al. [98–102] also confirmed that the dehydration of hydrous minerals will release a large amount of the water-bearing fluids, which can be used to reasonably explain the anomalously seismic and electrical transport behaviors in the region of hot subduction of deep mantle wedges.

4.3. Electrical Conductivity of Hydrous Amphibole

Global magnetotelluric (MT) results confirmed that the phenomenon of anomalously high electrical conductivity with its correspondent values of 10^{-2}–1 S/m is widely present in the regions of stable mid-lower Earth crust, the forearc of subduction zone, as well as the back arc of subduction zone. To interpret for the regionally anomalously high conductivity, several causes of formation models have already been presented, such as the salinity-bearing (or water-bearing) fluid, partial melting, graphite film in the grain boundary, as well as the impurity of high conductive mineral. Among them, the salinity-bearing (or water-bearing) fluid, as the most potential candidate, is mainly originated the released fluid phases during the process of the dehydration of hydrous minerals at high temperatures and pressures. Amphibole is of the most representative chain-structural silicate minerals, which can be stably existed the continental mid-crust region with its volume content of mineralogical composition up to 35–40% at the depth ranges from 15 to 30 km. Likewise, it is one of the crucial carriers for water transport with the volume percentage of highest abundance up to 50% outcropped at some epidote-amphibolite and amphibolite facies of the subduction zone [115].

Previous available investigations on the high-temperature ^{57}Fe Mossbauer spectra for the Fe-bearing calcic amphiboles observed that the amphibole dehydration occurred at the temperature ranges from 573 to 1073 K, which was accompanied by the loss of molecular structural hydroxyl as a form of either the hydrogen gas or molecular water [116]. Some researchers think that the high-temperature dehydration-induced hydrogen gas in Fe-bearing amphibole is produced at atmospheric pressure condition by virtue of the occurrence of the oxidation–dehydrogenation reaction (i.e., the electron jump between the ferrous ion and neighboring hydroxyl in the crystalline structure of hydrous amphibole),

$$2Fe^{2+}+2(OH)^{-} \rightarrow 2Fe^{3+}+2O^{2-}+H_2 \quad (7)$$

However, on the contrary, other researchers think that the high-temperature dehydration in Fe-bearing amphibole was resulted in the molecular water at atmospheric pressure condition originated from the heating in an oxidizing atmosphere results in a reaction at the sample surface, i.e., the dehydrogenation reaction,

$$2\left[Fe^{2+}+(OH)^{-}\right] + O_2 \rightarrow 4\left[Fe^{3+}+O^{2-}\right]+2H_2O \quad (8)$$

In order to systematically reveal the dehydration process of hydrous mineral, the electrical conductivity of iron-bearing amphibole was performed by Hu et al. (2018) at the temperature ranges from 623 to 1173 K and pressure ranges from 0.5 to 2.0 GPa within the frequency ranges of 10^{-1} to 10^{6} Hz, as illustrated in detail in Figure 8 [34]. When the experimental temperature is lower than 873 K, a linear increase between the logarithmic electrical conductivity of iron-bearing amphibole and temperature was observed. Subsequently, one moderate electrical conductivity of sample with pressure is confirmed, and an available discontinuous change of activation enthalpy before and after the dehydration reaction at the temperature around 873 K. Combined with the FT-IR measurements on the initial samples and recovered products, the dehydration temperature points of hydrous amphibole were verified as 843 ± 20 K. Thus, the electrical transport conduction of sample for the oxidation–dehydrogenation behavior was characterized by the variation of activation enthalpy in the Arrhenius equation before and after the dehydration reaction at high temperatures and pressures.

On the other hand, in comprehensive considerations of a relatively feeble discontinuity between the electrical conductivity of iron-bearing amphibole along with the temperature as well as some microscopic observations from the FT-IR measurement, optical microscope, and scanning electronic microscopy results before and after the electrical conductivity experiments, it is the first time that Hu et al. (2018) put forward that the dehydration reaction belongs to a oxidation–dehydrogenation process according to our obtained *in-situ* high-pressure electrical conductivity results [34]. Resembling previous temperature-induced dehydrogenation at the atmospheric pressure [117], the occurrence of oxidation–dehydrogenation reaction in hydrous amphibole can be described as,

$$Fe^{2+} + OH^{-} = Fe^{3+} + O^{2-} + \frac{1}{2}H_2 \quad (9)$$

In light of the oxidation–dehydrogenation reaction of Equation (9), the high-temperature dehydration-induced hydrogen gas in Fe-bearing amphibole was produced by the electron jump between the ferrous ion and neighboring hydroxyl in the crystalline structure of sample rather than the loss of molecular water at high pressure. Comparing these presently acquired measurement results, four previous high-pressure electrical conductivity data for iron-bearing amphibole from Glover and Vine (1994) [118]; Schmidbauer et al. (2000) [116]; Zhou et al. (2011) [119]; and Wang et al. (2012) [120] are also listed in Figure 8. Obviously, it makes clear that the electrical conductivity of hydrous Fe-bearing amphibole will increase with the rise of total iron content at HP-HT conditions (Figure 9). It is also observed that some available discrepancies between the previous and present electrical conductivity results may be originated from the different iron contents of Fe-bearing hydrous amphibole during the process of the occurrence of oxidation–dehydrogenation at high temperature and pressure.

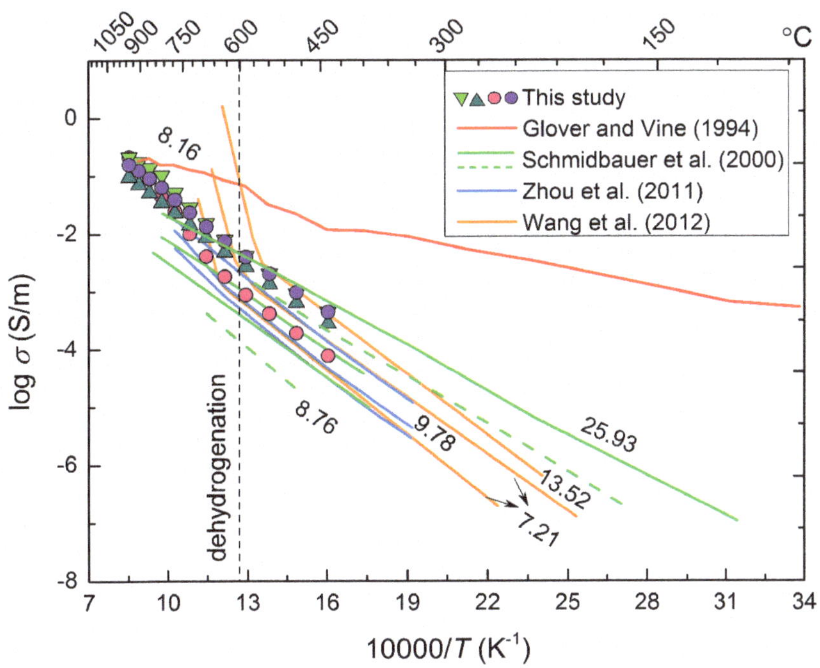

Figure 8. Comparisons between the acquired electrical conductivity of Fe-bearing hydrous amphibole electrical conductivity of amphibole and four previous high-pressure electrical conductivity data for iron-bearing amphibole from Glover and Vine (1994) [118]; Schmidbauer et al. (2000) [116]; Zhou et al. (2011) [119]; and Wang et al. (2012) [120]. The numbers represent the weight percentage (wt %) total ferrous and ferric iron in amphibole (reproduced with permission from the reference of Hu et al. [34]; published by Elsevier, 2018).

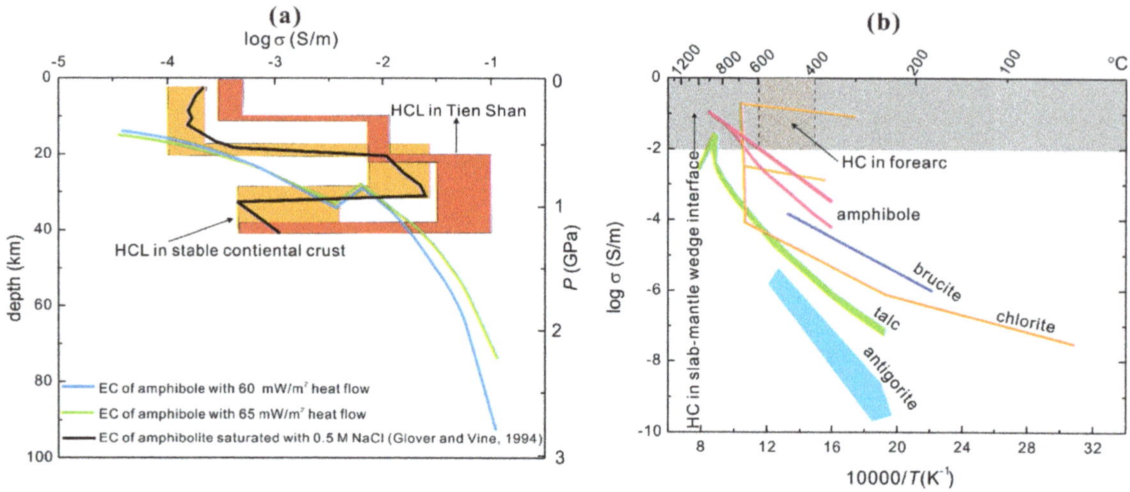

Figure 9. The electrical conductivity–depth profile according to these presently obtained electrical conductivity results on the Fe-bearing hydrous amphibole within the depth ranges from 20 to 40 km (**a**), and at the same time, several electrical conductivity results of some dominant hydrous minerals (e.g., NaCl-bearing saturated amphibole, antigorite, talc, brucite, and chlorite) in the subduction zone are also comprehensively compared (**b**). Data resource: the electrical conductivity of saturated amphibole with 0.5 sodium chloride from Glover and Vine (1994) [118], antigorite from Reynard et al. (2011) [121]; talc from Wang and Karato (2013) [122]; brucite from Guo and Yoshino (2014) [123], and chlorite from Manthilake et al. (2016) [101] (reproduced with permission from the reference of Hu et al. [34]; published by Elsevier, 2018).

4.4. Electrical Conductivity of Hydrous Kaolinite

As a typical aluminosilicate clay mineral, the hydrous kaolinite (its correspondent chemical molecular formula can be expressed as $Al_2Si_2O_5(OH)_4$) is widespread outcropped in the surface of Earth, as well as many different geological bodies and deep geodynamic processes, such as the kaolinization of hydrothermal mineral deposit, the natural alternation of representative rock-forming mineral (e.g., feldspar, et al.) and Meso-Cenozoic sedimentation basins. In the oceanic sedimentary products, kaolinite is one of the most important constituent compositions of water-bearing mineral, and its highest mineralogical content can reach up to 60%, which is possibly carrying a large amount of water to migrate in the subduction zone and resulting in the water cycle of the Earth interior [124,125]. Therefore, a systematic investigation on the physiochemical properties on the natural hydrous kaolinite at HP-HT conditions are crucial to deeply disclose its phase structure, storage state, and migration mechanism in the deep Earth interior.

Recently, Hong et al. (2022) performed the electrical conductivity and Raman spectroscopy experiments on natural hydrous kaolinite and its high-pressure polymorphs in order to explore the high-pressure phase structure and its phase stability during the process of compression and decompression under conditions of temperature ranges of 293–673 K and pressure ranges of 0.4–10.0 GPa [73]. By virtue of diamond anvil cell, the detailed experimental procedure and measurement method for the *in-situ* high-pressure electrical transport, structural and vibrational characterizations are previously displayed in the reference of Dai et al. (2019) [72]. Based on the acquired HP-HT electrical conductivity and Raman scattering results, the phase diagram was established at conditions of 2.3–6.5 GPa and 293–673 K within a depth range of 38–200 km, as detailed illustrated in Figure 10.

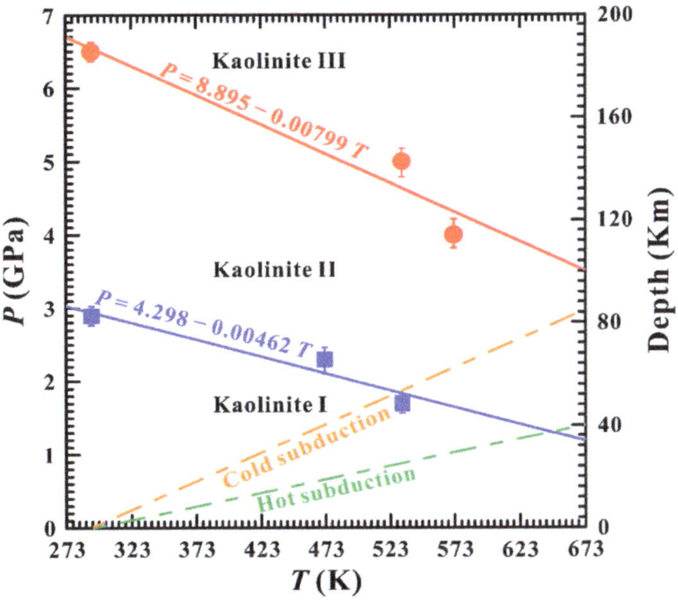

Figure 10. The temperature–pressure phase boundary based on the acquired high-pressure electrical conductivity and Raman scattering results on natural hydrous kaolinite at conditions of pressure ranges of 2.3–6.5 GPa, temperature ranges of 293–673 K, and depth ranges of 38–200 km. In here, two representative geothermal gradients of 5 and 10 K/km are selected to model the cold and hot subduction, respectively (reproduced with permission from the reference of Hong et al. [73]; published by Geoscience World, 2022).

In this present study, the initial sample is nominated as the kaolinite I phase at ambient conditions. With increasing pressure up to 2.9 GPa, another new phase of kaolinite II appeared, and subsequently, the phase transformation from kaolinite II to kaolinite III phases occurred at the critical pressure point of 6.5 GPa at atmospheric temperature. Upon decompression, a direct phase transformation of sample from kaolinite III to kaolinite I phases was occurred at the pressure point of 0.8 GPa rather than a continuous phase transition from kaolinite III to kaolinite II to kaolinite I phases. An obvious hysteresis effect of the structural phase transformation is also observed during the decompressed process. All of these available structural phase transformations in hydrous kaolinite upon compression and decompression were well characterized by the pressure-induced from those obviously discontinuous variations in the electrical conductivity and Raman shift at room temperature.

High-temperature and high pressure electrical conductivity and Raman spectroscopy results also confirmed that the critical pressure points of structural phase transition for natural hydrous kaolinite are strongly dependent on the experimental temperature. With increasing temperature, the pressure points of structural phase transformation for natural hydrous kaolinite gradually reduced. Further, it is the first time that the temperature–pressure phase boundary based on the acquired high-pressure electrical conductivity and Raman scattering results on natural hydrous kaolinite was successfully constructed at conditions of pressure ranges of 2.3–6.5 GPa, temperature ranges of 293–673 K, and depth ranges of 38–200 km. In addition, the fitted results for the structural phase transitions between the respective kaolinite I–II and kaolinite II–III phase boundaries can be expressed as,

$$P\ (\text{GPa}) = 4.298 - 0.00462T\ (\text{K}) \tag{10}$$

$$P\ (\text{GPa}) = 8.895 - 0.00799T\ (\text{K}) \quad (11)$$

By virtue of the obtained phase diagram, the high-pressure phase structure and its stability for natural hydrous kaolinite and its high-pressure polymorphs was well extrapolated by the laboratory-based electrical conductivity and Raman spectroscopy results within the depth range of 200 km under HP-HT conditions. Two representative geothermal gradients of 5 and 10 K/km are selected to model the corresponding cold and hot subduction environments, and it can provide a crucial constraint on the high-pressure behaviors of other kaolin-group minerals.

5. Concluding Remarks

In the past of several decades, with the rapid development of experimental techniques in the electrical conductivity measurement and high-pressure equipment, there are more and more electrical conductivity results to be reported on the hydrous silicate minerals in the Earth crust, upper mantle, and subduction zone at conditions of high temperatures and high pressures. In particular, some recent productions paid more attention on the investigations on electrical conductivity of nominally anhydrous and hydrous minerals at high temperature and high pressure. Two dominant presence species of molecular water and structural hydroxyl influence on the electrical conductivity of hydrous minerals are of the research point in the field of high-pressure mineral physics. In conjunction with those acquired experimental results from the laboratory-based high-pressure electrical conductivity, first-principle theoretical calculations and the observation of the field experimental rock, it can be efficiently applied to disclose the reason of high electrical conductivity from the global and regional field observations of magnetotelluric and geomagnetic deep sounding data.

For the main conduction mechanism, several typical electrical transport models occupying the position of crystalline lattice, such as hydrogen-related defects, iron-related defects, titanium-related defects, alkali-related defects, vacancy of interstitial ion, etc. have been already put forward on the base of high-temperature and high-pressure electrical conductivity on the hydrous silicate minerals. Owing to the presence of these complicated intrinsic crystalline defects, the dependence of the electrical conductivity of hydrous silicate minerals on the variations of influential factors (e.g., temperature, pressure, oxygen fugacity, structural phase transition, water content, dehydration effect, oxidation–dehydrogenation effect, crystallographic orientation, trace element of titanium content, iron content, alkali ion content, etc.) are overall working out. On the other hand, some external influential factors also possibly play a crucial role on the electrical conductivity of hydrous silicate minerals under conditions of high temperatures and high pressures, such as graphite layer of grain boundary, the impurity of magnetite, the impurity of chromite, the presence of some sulfide-bearing compounds, as well as salinity-bearing (or water-bearing) fluids in hydrous minerals. In addition, more and more attention has been paid to the electrical conductivity of hydrous minerals in the lower mantle, core-mantle boundary and Earth core using the diamond anvil cell in order to disclose the water cycling in the deep Earth interior.

In summary, in order to explore high conductivity anomaly in the deep Earth interior, we need to comprehensively consider all of these potential influential factors on the electrical conductivity of hydrous minerals at high temperature and high pressure, as well as geological background of a really regionally geotectonic unit. For example, just as pointed out by Dai and Karato (2020) [9], when we applied the laboratory-based electrical conductivity results from titanium-rich hydrous upper-mantle minerals to the titanium-poor real Earth (The titanium content is only ~50 ppm wt% in the upper mantle), it needs to be made with great care.

Author Contributions: All the authors contributed equally to this review paper. H.H. and L.D. designed the project. H.H., L.D., W.S., Y.Z., K.L., L.Y., C.P., M.H., M.W., Z.H., C.J., C.L., C.Y. and S.P. wrote the initial draft of the work, interpreted the results and the final version. All authors discussed the results and commented on the manuscript. All authors have read and agreed to the published version of the manuscript.

Funding: This research was funded by the NSF of China (grant numbers 42072055, 41774099 and 41772042), Youth Innovation Promotion Association of CAS (grant number 2019390), and Special Fund of the West Light Foundation of CAS.

Acknowledgments: Almost all listed original high-pressure electrical conductivity data were already obtained by Lidong Dai and his collaborators at the Chinese Academy of Sciences and Yale University. Here, please accept my honest thanks and greetings to Heping Li at the Key Laboratory of High-Temperature and High-Pressure Study of the Earth's Interior (HTHPSEI), Institute of Geochemistry, Chinese Academy of Sciences, the People's Republic of China, and Shun-ichiro Karato in the Karato High-pressure Laboratory, Department of Earth and Planetary Sciences, Yale University, United States.

Conflicts of Interest: The authors declare no conflict of interest.

References

1. Zheng, Y.; Hermann, J. Geochemistry of continental subduction-zone fluids. *Earth Planets Space* **2014**, *66*, 93. [CrossRef]
2. Karato, S. The role of hydrogen in the electrical conductivity of the upper mantle. *Nature* **1990**, *347*, 272–273. [CrossRef]
3. Selway, K.; O'Donnell, J.; Özaydin, S. Upper mantle melt distribution from petrologically constrained magnetotellurics. *Geochem. Geophys. Geosyst.* **2019**, *20*, 3328–3346. [CrossRef]
4. Selway, K. Negligible effect of hydrogen content on plate strength in East Africa. *Nat. Geosci.* **2015**, *8*, 543–546. [CrossRef]
5. Selway, K. On the causes of electrical conductivity anomalies in tectonically stable lithosphere. *Sur. Geophys.* **2014**, *35*, 219–257. [CrossRef]
6. Selway, K.; Hand, M.; Heinson, G.; Payne, J. Magnetotelluric constraints on subduction polarity: Reversing reconstruction models for Proterozoic Australia. *Geology* **2009**, *37*, 799–802. [CrossRef]
7. Förster, M.; Selway, K. Melting of subducted sediments reconciles geophysical images of subduction zones. *Nat. Commun.* **2021**, *12*, 1320. [CrossRef]
8. Yang, B.; Egbert, G.; Zhang, H.; Meqbel, N.; Hu, X. Electrical resistivity imaging of continental United States from three-dimensional inversion of EarthScope USArray magnetotelluric data. *Earth Planet. Sci. Lett.* **2021**, *576*, 117244. [CrossRef]
9. Dai, L.; Karato, S. Electrical conductivity of Ti-bearing hydrous olivine aggregates at high temperature and high pressure. *J. Geophys. Res. Solid Earth* **2020**, *125*, e2020JB020309. [CrossRef]
10. Liu, H.; Zhu, Q.; Xu, X.; Fei, H.; Yang, X. High electrical conductivity of olivine at oxidizing conditions of the shallow mantle and geophysical implications. *J. Geophys. Res. Solid Earth* **2021**, *126*, e2021JB022739. [CrossRef]
11. Sun, C.; Tang, G.; Fortin, J.; Borgomano, J.V.; Wang, S. Dispersion and attenuation of elastic wave velocities: Impact of microstructure heterogeneity and local measurements. *J. Geophys. Res. Solid Earth* **2020**, *125*, e2020JB020132. [CrossRef]
12. Horikawa, T.; Katsura, M.; Yokota, T.; Nakashima, S. Effects of pore water distributions on P-wave velocity–water saturation relations in partially saturated sandstones. *Geophys. J. Int.* **2021**, *226*, 1558–1573. [CrossRef]
13. Mohiuddin, A.; Karato, S.; Girard, J. Slab weakening during the olivine to ringwoodite transition in the mantle. *Nat. Geosci.* **2020**, *13*, 170–174. [CrossRef]
14. Zhang, Z.; Karato, S. The effect of pressure on grain-growth kinetics in olivine aggregates with some geophysical applications. *J. Geophys. Res. Solid Earth* **2021**, *126*, e2020JB020886. [CrossRef]
15. Sun, W.; Yoshino, T.; Sakamoto, N.; Yurimoto, H. Supercritical fluid in the mantle transition zone deduced from H–D interdiffusion of wadsleyite. *Earth Planet. Sci. Lett.* **2018**, *484*, 309–317. [CrossRef]
16. Fei, H.; Koizumi, S.; Sakamoto, N.; Hashiguchi, M.; Yurimoto, H.; Marquardt, K.; Miyajima, N.; Katsura, T. Pressure, temperature, water content, and oxygen fugacity dependence of the Mg grain-boundary diffusion coefficient in forsterite. *Am. Mineral.* **2018**, *103*, 1354–1361. [CrossRef]
17. Tielke, J.; Mecklenburgh, J.; Mariani, E.; Wheeler, J. The influence of water on the strength of olivine dislocation slip systems. *J. Geophys. Res. Solid Earth* **2019**, *124*, 6542–6559. [CrossRef]
18. Zhang, G.; Mei, S.; Song, M. Effect of water on the dislocation creep of enstatite aggregates at 300 MPa. *Geophys. Res. Lett.* **2020**, *47*, e2019GL085895. [CrossRef]
19. Précigout, J.; Prigent, C.; Palasse, L.; Pochon, A. Water pumping in mantle shear zones. *Nat. Commun.* **2017**, *8*, 15736. [CrossRef]
20. Speciale, P.; Behr, W.; Hirth, G.; Tokle, L. Rates of olivine grain growth during dynamic recrystallization and postdeformation annealing. *J. Geophys. Res. Solid Earth* **2020**, *125*, e2020JB020415. [CrossRef]
21. Huang, X.; Xu, Y.; Karato, S. Water content in the transition zone from electrical conductivity of wadsleyite and ringwoodite. *Nature* **2005**, *434*, 746–749. [CrossRef] [PubMed]

22. Huang, X.; Bai, W.; Xu, Y.; Karato, S. Influence of hydrogen on electrical conductivity of wadsleyite and ringwoodite with its geodynamics implications. *Acta Petrol. Sin.* **2005**, *21*, 1743–1748.
23. Huang, X.; Wang, X.; Chen, Z.; Bai, W. The experimental research on electrical conductivity for minerals and rocks under condition of upper mantle. *Sci. Sin. Terrae* **2017**, *47*, 518–529.
24. Wang, D.; Mookherjee, M.; Xu, Y.; Karato, S. The effect of water on the electrical conductivity of olivine. *Nature* **2006**, *443*, 977–980. [CrossRef]
25. Dai, L.; Karato, S. Electrical conductivity of wadsleyite at high temperatures and high pressures. *Earth Planet. Sci. Lett.* **2009**, *287*, 277–283. [CrossRef]
26. Dai, L.; Karato, S. Electrical conductivity of pyrope-rich garnet at high temperature and high pressure. *Phys. Earth Planet. Inter.* **2009**, *176*, 83–88. [CrossRef]
27. Dai, L.; Karato, S. Electrical conductivity of orthopyroxene: Implications for the water content of the asthenosphere. *Proc. Jpn. Acad.* **2009**, *85*, 466–475. [CrossRef]
28. Dai, L.; Karato, S. High and highly anisotropic electrical conductivity of the asthenosphere due to hydrogen diffusion in olivine. *Earth Planet. Sci Lett.* **2014**, *408*, 79–86. [CrossRef]
29. Dai, L.; Karato, S. The effect of pressure on the electrical conductivity of olivine under the hydrogen-rich conditions. *Phys. Earth Planet. Inter.* **2014**, *232*, 51–56. [CrossRef]
30. Dai, L.; Karato, S. Influence of oxygen fugacity on the electrical conductivity of hydrous olivine: Implications for the mechanism of conduction. *Phys. Earth Planet. Inter.* **2014**, *232*, 57–60. [CrossRef]
31. Dai, L.; Karato, S. Influence of FeO and H on the electrical conductivity of olivine. *Phys. Earth Planet. Inter.* **2014**, *237*, 73–79. [CrossRef]
32. Karato, S.; Dai, L. Comments on "Electrical conductivity of wadsleyite as a function of temperature and water content" by Manthilake et al. *Phys. Earth Planet. Inter.* **2009**, *174*, 19–21. [CrossRef]
33. Hu, H.; Dai, L.; Li, H.; Hui, K.; Sun, W. Influence of dehydration on the electrical conductivity of epidote and implications for high-conductivity anomalies in subduction zones. *J. Geophys. Res. Solid Earth* **2017**, *122*, 2751–2762. [CrossRef]
34. Hu, H.; Dai, L.; Li, H.; Sun, W.; Li, B. Effect of dehydrogenation on the electrical conductivity of Fe-bearing amphibole: Implications for high conductivity anomalies in subduction zones and continental crust. *Earth Planet. Sci Lett.* **2018**, *498*, 27–37. [CrossRef]
35. Dai, L.; Hu, H.; Jiang, J.; Sun, W.; Li, H.; Wang, M.; Vallianatos, F.; Saltas, V. An overview of the experimental studies on the electrical conductivity of major minerals in the upper mantle and transition zone. *Materials* **2020**, *13*, 408. [CrossRef]
36. He, Y.; Dai, L.; Kim, D.Y.; Li, H.; Karato, S. Thermal ionization of hydrogen in hydrous olivine with enhanced and anisotropic conductivity. *J. Geophys. Res. Solid Earth* **2021**, *126*, e2021JB022939. [CrossRef]
37. Shan, S.; Xiao, C.; Li, H.; Xu, L.; Lin, S.; Li, S. In Situ electrical conductivity measurements of porous water-containing rock materials under high temperature and high pressure conditions in an autoclave. *Rev. Sci. Instrum.* **2021**, *92*, 095104. [CrossRef]
38. Dai, L.; Li, H.; Hu, H.; Shan, S. Novel technique to control oxygen fugacity during high-pressure measurements of grain boundary conductivities of rocks. *Rev. Sci. Instrum.* **2009**, *80*, 033903. [CrossRef]
39. Gwanmesia, G.; Whitaker, M.; Dai, L.; James, A.; Chen, H.; Triplett, R.; Cai, N. The elastic properties of β-Mg_2SiO_4 containing 0.73 wt.% of H_2O to 10 GPa and 600 K by ultrasonic interferometry with synchrotron x-radiation. *Minerals* **2020**, *10*, 209. [CrossRef]
40. Paterson, M. The determination of hydroxyl by infrared absorption in quartz, silicate glass and similar materials. *Bull. De Minéralogie* **1982**, *105*, 20–29. [CrossRef]
41. Dai, L.; Li, H.; Hu, H.; Shan, S.; Jiang, J.; Hui, K. The effect of chemical composition and oxygen fugacity on the electrical conductivity of dry and hydrous garnet at high temperatures and pressures. *Contrib. Mineral. Petrol.* **2012**, *163*, 689–700. [CrossRef]
42. Dai, L.; Hu, H.; Li, H.; Hui, K.; Jiang, J.; Li, J.; Sun, W. Electrical conductivity of gabbro: The effects of temperature, pressure and oxygen fugacity. *Eur. J. Mineral.* **2015**, *27*, 215–224. [CrossRef]
43. Dai, L.; Jiang, J.; Li, H.; Hu, H.; Hui, K. Electrical conductivity of hydrous natural basalt at high temperatures and high pressures. *J. Appl. Geophys.* **2015**, *112*, 290–297. [CrossRef]
44. Sun, W.; Dai, L.; Li, H.; Hu, H.; Wu, L.; Jiang, J. Electrical conductivity of mudstone (before and after dehydration at high PT) and a test of high conductivity layers in the crust. *Am. Mineral.* **2017**, *102*, 2450–2456. [CrossRef]
45. Sun, W.; Dai, L.; Li, H.; Hu, H.; Jiang, J.; Hui, K. Effect of dehydration on the electrical conductivity of phyllite at high temperatures and pressures. *Mineral. Petrol.* **2017**, *111*, 853–863. [CrossRef]
46. Dai, L.; Hu, H.; Sun, W.; Li, H.; Liu, C.; Wang, M. Influence of high conductive magnetite impurity on the electrical conductivity of dry olivine aggregates at high temperature and high pressure. *Minerals* **2019**, *9*, 44. [CrossRef]
47. Dai, L.; Sun, W.; Li, H.; Hu, H.; Wu, L.; Jiang, J. Effect of chemical composition on the electrical conductivity of gneiss at high temperatures and pressures. *Solid Earth* **2018**, *9*, 233–245. [CrossRef]
48. Dai, L.; Hu, H.; Li, H.; Sun, W.; Jiang, J. Influence of anisotropy on the electrical conductivity and diffusion coefficient of dry K-feldspar: Implications for the mechanism of conduction. *Chin. Phys. B* **2018**, *27*, 028703. [CrossRef]
49. Dai, L.; Hu, H.; Li, H.; Wu, L.; Hui, K.; Jiang, J.; Sun, W. Influence of temperature, pressure, and oxygen fugacity on the electrical conductivity of dry eclogite, and geophysical implications. *Geochem. Geophys. Geosyst.* **2016**, *17*, 2394–2407. [CrossRef]

50. Dai, L.; Hu, H.; Li, H.; Jiang, J.; Hui, K. Influence of temperature, pressure, and chemical composition on the electrical conductivity of granite. *Am. Mineral.* **2014**, *99*, 1420–1428. [CrossRef]
51. Dai, L.; Li, H.; Hu, H.; Jiang, J.; Hui, K.; Shan, S. Electrical conductivity of $Alm_{82}Py_{15}Grs_{3}$ almandine-rich garnet determined by impedance spectroscopy at high temperatures and high pressures. *Tectonophysics* **2013**, *608*, 1086–1093. [CrossRef]
52. Dai, L.; Li, H.; Hu, H.; Shan, S. In-Situ control of oxygen fugacity for laboratory measurements of electrical conductivity of minerals and rocks in multi-anvil press. *Chin. Phys. B* **2011**, *20*, 049101. [CrossRef]
53. Dai, L.; Li, H.; Li, C.; Hu, H.; Shan, S. The electrical conductivity of dry polycrystalline olivine compacts at high temperatures and pressures. *Mineral. Mag.* **2010**, *74*, 849–857. [CrossRef]
54. Dai, L.; Li, H.; Hu, H.; Shan, S. Experimental study of grain boundary electrical conductivities of dry synthetic peridotite under high-temperature, high-pressure, and different oxygen fugacity conditions. *J. Geophys. Res. Solid Earth* **2008**, *113*, B12211. [CrossRef]
55. Dai, L.; Li, H.; Deng, H.; Liu, C.; Su, G.; Shan, S.; Zhang, L.; Wang, R. In-Situ control of different oxygen fugacity experimental study on the electrical conductivity of lherzolite at high temperature and high pressure. *J. Phys. Chem. Solids* **2008**, *69*, 101–110. [CrossRef]
56. Dai, L.; Li, H.; Liu, C.; Su, G.; Shan, S. Experimental measurement on the electrical conductivity of pyroxenite at high temperature and high pressure under different oxygen fugacities. *High Press. Res.* **2006**, *26*, 193–202. [CrossRef]
57. Dai, L.; Li, H.; Liu, C.; Cui, T.; Shan, S.; Yang, C.; Liu, Q.; Deng, H. Experimental measurement on the electrical conductivity of single crystal olivine at high temperature and high pressure under different oxygen fugacities. *Prog. Nat. Sci.* **2006**, *16*, 387–393. [CrossRef]
58. Dai, L.; Li, H.; Liu, C.; Shan, S.; Cui, T.; Su, G. Experimental study on the electrical conductivity of orthopyroxene at high temperature and high pressure under different oxygen fugacities. *Acta Geol. Sin. Engl.* **2005**, *79*, 803–809.
59. Dai, L.; Li, H.; Liu, C.; Su, G.; Cui, T. In Situ control of oxygen fugacity experimental study on the crystallographic anisotropy of the electrical conductivities of diopside at high temperature and high pressure. *Acta Petrol. Sin.* **2005**, *21*, 1737–1742.
60. Hu, H.; Dai, L.; Li, H.; Hui, K.; Li, J. Temperature and pressure dependence of electrical conductivity in synthetic anorthite. *Solid State Ion.* **2015**, *276*, 136–141. [CrossRef]
61. Hu, H.; Dai, L.; Li, H.; Jiang, J.; Hui, K. Electrical conductivity of K-feldspar at high temperature and high pressure. *Mineral. Petrol.* **2014**, *108*, 609–618. [CrossRef]
62. Hu, H.; Li, H.; Dai, L.; Shan, S.; Zhu, C. Electrical conductivity of alkali feldspar solid solutions at high temperatures and high pressures. *Phys. Chem. Mineral.* **2013**, *40*, 51–62. [CrossRef]
63. Hu, H.; Li, H.; Dai, L.; Shan, S.; Zhu, C. Electrical conductivity of albite at high temperatures and high pressures. *Am. Miner.* **2011**, *96*, 1821–1827. [CrossRef]
64. Hui, K.; Dai, L.; Li, H.; Hu, H.; Jiang, J.; Sun, W.; Zhang, H. Experimental study on the electrical conductivity of pyroxene andesite at high temperature and high pressure. *Pure Appl. Geophys.* **2017**, *174*, 1033–1041. [CrossRef]
65. Hui, K.; Zhang, H.; Li, H.; Dai, L.; Hu, H.; Jiang, J.; Sun, W. Experimental study on the electrical conductivity of quartz andesite at high temperature and high pressure: Evidence of grain boundary transport. *Solid Earth* **2015**, *6*, 1037–1043. [CrossRef]
66. Sun, W.; Dai, L.; Hu, H.; Jiang, J.; Wang, M.; Hu, Z.; Jing, C. Influence of saline fluids on the electrical conductivity of olivine aggregates at high temperature and high pressure and its geological implications. *Front. Earth Sci.* **2021**, *9*, 749896. [CrossRef]
67. Sun, W.; Jiang, J.; Dai, L.; Hu, H.; Wang, M.; Qi, Y.; Li, H. Electrical properties of dry polycrystalline olivine mixed with various chromite contents: Implications for the high-conductivity anomalies in subduction zones. *Geosci. Front.* **2021**, *12*, 101178. [CrossRef]
68. Sun, W.; Dai, L.; Li, H.; Hu, H.; Jiang, J.; Wang, M. Electrical conductivity of clinopyroxene-NaCl-H_2O system at high temperatures and pressures: Implications for high-conductivity anomalies in the deep crust and subduction zone. *J. Geophys. Res. Solid Earth* **2020**, *125*, e2019JB019093. [CrossRef]
69. Sun, W.; Dai, L.; Li, H.; Hu, H.; Jiang, J.; Liu, C. Experimental study on the electrical properties of carbonaceous slate: A special natural rock with unusually high conductivity at high temperatures and pressures. *High Temp. High Press.* **2019**, *48*, 455–467. [CrossRef]
70. Sun, W.; Dai, L.; Li, H.; Hu, H.; Liu, C. Effect of temperature, pressure and chemical composition on the electrical conductivity of granulite and geophysical implications. *J. Mineral. Petrol. Sci.* **2019**, *114*, 87–98. [CrossRef]
71. Sun, W.; Dai, L.; Li, H.; Hu, H.; Liu, C.; Wang, M. Effect of temperature, pressure and chemical compositions on the electrical conductivity of schist: Implications for electrical structures under the Tibetan plateau. *Materials* **2019**, *12*, 961. [CrossRef] [PubMed]
72. Dai, L.; Pu, C.; Li, H.; Hu, H.; Liu, K.; Yang, L.; Hong, M. Characterization of metallization and amorphization for GaP under different hydrostatic environments in diamond anvil cell up to 40.0 GPa. *Rev. Sci. Instrum.* **2019**, *90*, 066103. [CrossRef] [PubMed]
73. Hong, M.; Dai, L.; Hu, H.; Yang, L.; Zhang, X. Pressure-induced structural phase transitions in natural kaolinite investigated by Raman spectroscopy and electrical conductivity. *Am. Mineral.* **2022**, *in press*. [CrossRef]
74. Liu, K.; Dai, L.; Li, H.; Hu, H.; Zhuang, Y.; Yang, L.; Pu, C.; Hong, M. Pressure-induced phase transitions for goethite investigated by Raman spectroscopy and electrical conductivity. *High Press. Res.* **2019**, *39*, 106–116. [CrossRef]
75. Pu, C.; Dai, L.; Li, H.; Hu, H.; Zhuang, Y.; Liu, K.; Yang, L.; Hong, M. High–pressure electrical conductivity and Raman spectroscopic study of chalcanthite. *Spectrosc. Lett.* **2018**, *51*, 531–539. [CrossRef]

76. Yang, L.; Dai, L.; Li, H.; Hu, H.; Hong, M.; Zhang, X. The phase transition and dehydration in epsomite under high temperature and high pressure. *Crystals* **2020**, *10*, 75. [CrossRef]
77. Yang, L.; Dai, L.; Li, H.; Hu, H.; Zhuang, Y.; Liu, K.; Pu, C.; Hong, M. Pressure-induced structural phase transition and dehydration for gypsum investigated by Raman spectroscopy and electrical conductivity. *Chem. Phys. Lett.* **2018**, *706*, 151–157. [CrossRef]
78. Dai, L.; Liu, K.; Li, H.; Wu, L.; Hu, H.; Zhuang, Y.; Yang, L.; Pu, C.; Liu, P. Pressure-induced irreversible metallization accompanying the phase transitions in Sb_2S_3. *Phys. Rev. B* **2018**, *97*, 024103. [CrossRef]
79. Dai, L.; Zhuang, Y.; Li, H.; Wu, L.; Hu, H.; Liu, K.; Yang, L.; Pu, C. Pressure-induced irreversible amorphization and metallization with a structural phase transition in arsenic telluride. *J. Mater. Chem. C* **2017**, *5*, 12157–12162. [CrossRef]
80. Dai, L.; Wu, L.; Li, H.; Hu, H.; Zhuang, Y.; Liu, K. Evidence of the pressure-induced conductivity switching of yttrium-doped $SrTiO_3$. *J. Phys Condens. Matt.* **2016**, *28*, 475501. [CrossRef]
81. Dai, L.; Wu, L.; Li, H.; Hu, H.; Zhuang, Y.; Liu, K. Pressure-induced phase-transition and improvement of the micro dielectric properties in yttrium-doped $SrZrO_3$. *Europhys. Lett.* **2016**, *114*, 56003. [CrossRef]
82. Hong, M.; Dai, L.; Hu, H.; Zhang, X. Pressure-induced structural phase transition and metallization in Ga_2Se_3 up to 40.2 GPa under non-hydrostatic and hydrostatic environments. *Crystals* **2021**, *11*, 746. [CrossRef]
83. Hong, M.; Dai, L.; Li, H.; Hu, H.; Liu, K.; Yang, L.; Pu, C. Structural phase transition and metallization of nanocrystalline rutile investigated by high-pressure Raman spectroscopy and electrical conductivity. *Minerals* **2019**, *9*, 441. [CrossRef]
84. Liu, K.; Dai, L.; Li, H.; Hu, H.; Yang, L.; Pu, C.; Hong, M. Evidences for phase transition and metallization in β-In_2S_3 at high pressure. *Chem. Phys.* **2019**, *524*, 63–69. [CrossRef]
85. Liu, K.; Dai, L.; Li, H.; Hu, H.; Yang, L.; Pu, C.; Hong, M. Phase transition and metallization of orpiment by Raman spectroscopy, electrical conductivity and theoretical calculation under high pressure. *Materials* **2019**, *12*, 784. [CrossRef]
86. Liu, K.; Dai, L.; Li, H.; Hu, H.; Wu, L.; Zhuang, Y.; Pu, C.; Yang, L. Migration of impurity level reflected in the electrical conductivity variation for natural pyrite at high temperature and high pressure. *Phys. Chem. Miner.* **2018**, *45*, 85–92. [CrossRef]
87. Pu, C.; Dai, L.; Li, H.; Hu, H.; Liu, K.; Yang, L.; Hong, M. Pressure-induced phase transitions of ZnSe under different pressure environments. *AIP Adv.* **2019**, *9*, 025004. [CrossRef]
88. Wu, L.; Dai, L.; Li, H.; Hu, H.; Zhuang, Y.; Liu, K. Anomalous phase transition of Bi-doped Zn_2GeO_4 investigated by electrical conductivity and Raman spectroscopy under high pressure. *J. Appl. Phys.* **2017**, *121*, 125901. [CrossRef]
89. Wu, L.; Dai, L.; Li, H.; Zhuang, Y.; Liu, K. Pressure-induced improvement of grain boundary properties in Y-doped $BaZrO_3$. *J. Phys. D Appl. Phys.* **2016**, *49*, 345102. [CrossRef]
90. Yang, L.; Jiang, J.; Dai, L.; Hu, H.; Hong, M.; Zhang, X.; Li, H.; Liu, P. High-pressure structural phase transition and metallization in Ga_2S_3 under non-hydrostatic and hydrostatic conditions up to 36.4 GPa. *J. Mater. Chem. C* **2021**, *9*, 2912–2918. [CrossRef]
91. Yang, L.; Dai, L.; Li, H.; Hu, H.; Hong, M.; Zhang, X.; Liu, P. High-pressure investigations on the isostructural phase transition and metallization in realgar with diamond anvil cells. *Geosci. Front.* **2021**, *12*, 1031–1037. [CrossRef]
92. Yang, L.; Dai, L.; Li, H.; Hu, H.; Liu, K.; Pu, C.; Hong, M.; Liu, P. Characterization of the pressure-induced phase transition of metallization for MoTe2 under different hydrostatic environments. *AIP Adv.* **2019**, *9*, 065104. [CrossRef]
93. Yang, L.; Dai, L.; Li, H.; Hu, H.; Liu, K.; Pu, C.; Hong, M.; Liu, P. Pressure-induced metallization in $MoSe_2$ under different pressure conditions. *RSC Adv.* **2019**, *9*, 5794–5803. [CrossRef]
94. Zhang, X.; Dai, L.; Hu, H.; Hong, M. Pressure-induced metallic phase transition in gallium arsenide up to 24.3 GPa under hydrostatic conditions. *Mod. Phys. Lett. B* **2021**, *35*, 2150460. [CrossRef]
95. Zhuang, Y.; Dai, L.; Li, H.; Hu, H.; Liu, K.; Yang, L.; Pu, C.; Hong, M. Pressure induced reversible metallization and phase transition in Zinc Telluride. *Mod. Phys. Lett. B* **2018**, *34*, 1850342. [CrossRef]
96. Zhuang, Y.; Dai, L.; Li, H.; Hu, H.; Liu, K.; Yang, L.; Pu, C.; Hong, M.; Liu, P. Deviatoric stresses promoted metallization in rhenium disulfide. *J. Phys. D Appl. Phys.* **2018**, *51*, 165101. [CrossRef]
97. Zhuang, Y.; Dai, L.; Wu, L.; Li, H.; Hu, H.; Liu, K.; Yang, L.; Pu, C. Pressure-induced permanent metallization with reversible structural transition in molybdenum disulfide. *Appl. Phys. Lett.* **2017**, *110*, 122103. [CrossRef]
98. Manthilake, G.; Koga, K.; Peng, Y.; Mookherjee, M. Halogen bearing amphiboles, aqueous fluids, and melts in subduction zones: Insights on halogen cycle from electrical conductivity. *J. Geophys. Res. Solid Earth* **2021**, *126*, e2020JB021339. [CrossRef]
99. Manthilake, G.; Peng, Y.; Koga, K.; Mookherjee, M. Tracking slab surface temperatures with electrical conductivity of glaucophane. *Sci. Rep.* **2021**, *11*, 18014. [CrossRef]
100. Manthilake, G.; Schiavi, F.; Zhao, C.; Mookherjee, M.; Bouhifd, M.; Jouffret, L. The electrical conductivity of Liebermannite: Implications for water transport into the Earth's lower mantle. *J. Geophys. Res. Solid Earth* **2020**, *125*, e2020JB020094. [CrossRef]
101. Manthilake, G.; Bolfan-Casanova, N.; Novella, D.; Mookherjee, M.; Andrault, D. Dehydration of chlorite explains anomalously high electrical conductivity in the mantle wedges. *Sci. Adv.* **2016**, *2*, e1501631. [CrossRef] [PubMed]
102. Manthilake, G.; Mookherjee, M.; Bolfan-Casanova, N.; Andrault, D. Electrical conductivity of lawsonite and dehydrating fluids at high pressures and temperatures. *Geophys. Res. Lett.* **2015**, *42*, 7398–7405. [CrossRef]
103. Saltas, V.; Pentari, D.; Vallianatos, F. Complex electrical conductivity of biotite and muscovite micas at elevated temperatures: A comparative study. *Materials* **2020**, *13*, 3513. [CrossRef]
104. Saltas, V.; Fitilis, I.; Vallianatos, F. A combined complex electrical impedance and acoustic emission study in limestone samples under uniaxial loading. *Tectonophysics* **2014**, *637*, 198–206. [CrossRef]

105. Saltas, V.; Chatzistamou, V.; Pentari, D.; Paris, E.; Triantis, D.; Fitilis, I.; Vallianatos, F. Complex electrical conductivity measurements of a KTB amphibolite sample at elevated temperatures. *Mater. Chem. Phys.* **2013**, *139*, 169–175. [CrossRef]
106. Saltas, V.; Vallianatos, F.; Gidarakos, E. Charge transport in diatomaceous earth studied by broadband dielectric spectroscopy. *Appl. Clay Sci.* **2013**, *80–81*, 226–235. [CrossRef]
107. Huebner, J.; Dillenburg, R. Impedance spectra of hot, dry silicate minerals and rock: Qualitative interpretation of spectra. *Am. Mineral.* **1995**, *80*, 46–64. [CrossRef]
108. Macdonald, J.; Barsoukov, E. *Impedance Spectroscopy: Theory, Experiment, and Applications*, 2nd ed.; Wiley-Interscience: New York, NY, USA, 2005; pp. 1–595.
109. Ter Heege, J.; Renner, J. In situ impedance spectroscopy on pyrophyllite and $CaCO_3$ at high pressure and temperature: Phase transformations and kinetics of atomistic transport. *Phys. Chem. Miner.* **2007**, *34*, 445–465. [CrossRef]
110. Cline, C., II; Faul, U.; David, E.; Berry, A.; Jackson, I. Redox-influenced seismic properties of upper-mantle olivine. *Nature* **2018**, *555*, 355–358. [CrossRef]
111. Tollan, P.; Smith, R.; O'Neill, H.; Hermann, J. The responses of four main substitution mechanisms of H in olivine to H_2O activity at 1050 °C and 3 GPa. *Prog. Earth Planet. Sci.* **2017**, *4*, 14. [CrossRef]
112. Poli, S.; Schmidt, M. Experimental subsolidus studies on epidote minerals. *Rev. Mineral. Geochem.* **2004**, *56*, 171–195. [CrossRef]
113. Bagdassarov, N.; Delépine, N. α–β inversion in quartz from low frequency electrical impedance spectroscopy. *J. Phys. Chem. Solids* **2004**, *65*, 1517–1526. [CrossRef]
114. Sinmyo, R.; Keppler, H. Electrical conductivity of NaCl-bearing aqueous fluids to 600 °C and 1 GPa. *Contrib. Mineral. Petrol.* **2017**, *172*, 4. [CrossRef]
115. Christensen, N.; Mooney, W. Seismic velocity structure and composition of the continental crust: A global view. *J. Geophys. Res. Solid Earth* **1995**, *100*, 9761–9788. [CrossRef]
116. Schmidbauer, E.; Kunzmann, T.; Fehr, T.; Hochleitner, R. Electrical resistivity and ^{57}Fe Mossbauer spectra of Fe-bearing calcic amphiboles. *Phys. Chem. Miner.* **2000**, *27*, 347–356. [CrossRef]
117. Phillips, M.; Popp, R.; Clowe, C. Structural adjustments accompanying oxidation-dehydrogenation in amphiboles. *Am. Mineral.* **1988**, *73*, 500–506.
118. Glover, P.; Vine, F. Electrical conductivity of the continental crust. *Geophys. Res. Lett.* **1994**, *21*, 2357–2360. [CrossRef]
119. Zhou, W.; Fan, D.; Liu, Y.; Xie, H. Measurements of wave velocity and electrical conductivity of an amphibolite from southwestern margin of the Tarim Basin at pressures to 1.0 GPa and temperatures to 700 °C: Comparison with field observations. *Geophys. J. Int.* **2011**, *187*, 1393–1404. [CrossRef]
120. Wang, D.; Guo, Y.; Yu, Y.; Karato, S. Electrical conductivity of amphibole-bearing rocks: Influence of dehydration. *Contrib. Mineral. Petrol.* **2012**, *164*, 17–25. [CrossRef]
121. Reynard, B.; Mibe, K.; de Moortèle, B.V. Electrical conductivity of the serpentinised mantle and fluid flow in subduction zones. *Earth Planet. Sci. Lett.* **2011**, *307*, 387–394. [CrossRef]
122. Wang, D.; Karato, S. Electrical conductivity of talc aggregates at 0.5 GPa: Influence of dehydration. *Phys. Chem. Miner.* **2013**, *40*, 11–17. [CrossRef]
123. Guo, X.; Yoshino, T. Pressure-induced enhancement of proton conduction in brucite. *Geophys. Res. Lett.* **2014**, *41*, 813–819. [CrossRef]
124. Balan, E.; Calas, G.; Bish, D. Kaolin-group minerals: From hydrogen-bonded layers to environmental recorders. *Elements* **2014**, *10*, 183–188. [CrossRef]
125. Schroeder, P.; Erickson, G. Kaolin: From ancient porcelains to nanocomposites. *Elements* **2014**, *10*, 177–182. [CrossRef]

minerals

MDPI

Article

Design and Application of a Rock Porosity Measurement Apparatus under High Isostatic Pressure

Liyu Liu [1,2], Heping Li [1,*], Hongbin Zhou [1], Sen Lin [1] and Shengbin Li [1]

[1] Key Laboratory of High-Temperature and High-Pressure Study of the Earth's Interior, Institute of Geochemistry, Chinese Academy of Sciences, Guiyang 550081, China; liuliyu@mail.gyig.ac.cn (L.L.); zhouhongbin@vip.gyig.ac.cn (H.Z.); linsen@mail.gyig.ac.cn (S.L.); lishengbin@mail.gyig.ac.cn (S.L.)
[2] College of Earth and Planetary Sciences, University of Chinese Academy of Sciences, Beijing 100049, China
* Correspondence: liheping@vip.gyig.ac.cn

Abstract: Rock porosity is a key physical parameter at room temperature and pressure that plays an important role in evaluating reserves of oil and natural gas. Research on rock porosity spans over a hundred years. However, in situ porosity under a high isostatic pressure has not been adequately explored, and the experimental conditions for measuring porosity remain unclear. To investigate the feasibility of porosity measurement under a high isostatic pressure and the optimal choice of experimental conditions for this, we design an experimental apparatus that can achieve isostatic pressure up to 200 MPa to fit the relationship between the void volume of a given sample and the drop in gas pressure in an empty standard chamber. The effect of experimental parameters, such as the initial gas pressure at the inlet, the time needed for the gas to reach equilibrium, and the time needed for vacuuming, on the porosity experiment was examined. A series of porosity experiments under different isostatic pressures of up to 200 MPa were carried out with this apparatus. The results quantitatively verify the degree to which porosity is related to isostatic pressure.

Keywords: rock porosity; high isostatic pressure; sample holder; sandstone

Citation: Liu, L.; Li, H.; Zhou, H.; Lin, S.; Li, S. Design and Application of a Rock Porosity Measurement Apparatus under High Isostatic Pressure. *Minerals* **2022**, *12*, 127. https://doi.org/10.3390/min12020127

Academic Editors: Lidong Dai, Haiying Hu, Jianjun Jiang and Rossella Arletti

Received: 16 December 2021
Accepted: 20 January 2022
Published: 21 January 2022

1. Introduction

Porosity is an essential parameter in the study of reservoir physics and solid earth physics because it provides empirical support for the evaluation of reserves of oil and natural gas as well as the interpretation of anomalies in the velocities of crustal waves [1–3]. In general, it is described as the ratio of the volume of voids in a given sample to the total volume of the sample, and is usually expressed as a percentage [4,5].

To obtain the porosity of reservoirs, some porosity-logging methods have been proposed, these methods evaluate porosity based on empirical data, such as continuous or interval velocity logging [6–8], Neutron logging [9], Density logging [10], and nuclear magnetic resonance (NMR) well logging [11]. However, geological structures are complex, and logging methods measure porosity fluctuations significantly and exhibit large errors. To better calibrate the logging analysis or investigate rock porosity more precisely, laboratory techniques for measuring porosity have been developed. The relevant methods are divided into two categories, one is the direct observation method, including X-ray computed tomography (CT) [12,13], core porosity measurement using gamma rays [14], and thin sections analysis [15]. The other is the experimental method, including mercury intrusion porosimetry (MIP) [16], the pressure/mass method [17], and the gas expansion method [18–20].

The direct observation methods could intuitively obtain the texture of interconnected and isolated rock voids. Unfortunately, only the type of interconnected voids was of value to our practical application. The information of isolated rock voids may interfere with the accuracy of interconnected pore statistics. Consequently, the experimental method has been commonly adopted to evaluate rock porosity. The MIP occurs through the injection of mercury into rock pores, which can be used to roughly estimate the porosity of rocks. However, rock voids in different reservoirs are different by genesis, pore morphology, and size. Rocks generally cannot be wetted by mercury. In order to inject mercury into the rock pore, the external pressure is applied. This operation often results in damage to the sample [18,21]. By contrast, the gas expansion method, which is based on Boyle's law, has been adopted as a more popular method for measuring porosity [22]. The measurement gas medium of this method usually includes air, nitrogen, argon, or helium. All of these gas mediums have a lower viscosity than that of liquid, and can enter certain fine pores that liquid is unable to penetrate [23,24].

Although it has been studied extensively at room temperature and ambient pressure, rock porosity at high confining pressure (equal on all sides) has not been adequately explored in research. In general, with the increase in confining pressure, rock voids decrease, some of them even close completely, which leads to the decrease in rock porosity

However, the existing experimental conditions under room temperature and ambient pressure cannot reflect the state of rock pores under high confining pressure. In addition, the existing experimental equipment had difficulty meeting the experimental requirements. Due to the thin wall thickness of the core holder and its poor sealing performance at high confining pressures, the core holder can hardly perform effectively at high confining pressures [25,26]. The available high-pressure pressurization device has difficulty providing the target confining pressure. Attempts have been made to determine rock porosity under confining pressures and temperature [27,28], the confining pressure generally used is not more than 90 MPa. Furthermore, due to the separated axial and radial pressures, the confining pressure belongs to non-isostatic pressure [29]. Thus, it is essential to solve the pressure resistance and the sealing performance of the core holder under high confining pressure in order to be able to measure rock porosity at a high confining pressure. We propose that the confining pressure on the sample in the experiment is isostatic pressure.

In this paper, we first built a porosity measurement device that was used for 200 MPa confining pressure by taking the autoclave in the laboratory as a sample holder. We also tested the porosity of sandstone at different confining pressures of up to 200 MPa. Meanwhile, some experimental parameters were preliminarily explored in the proposed apparatus.

2. Principle of Experiment

The common reservoir rocks include limestone, dolomite, sandstone, and shale, all of which consist of a matrix and a void. There are two types of void space, both interconnected and isolated. The effective porosity includes only voids that are interconnected [30]. The porosity measured by the proposed apparatus is the effective porosity. It is defined as

$$\varnothing_{eff} = \frac{V_{iv}}{V_b} \tag{1}$$

where $\varnothing_{eff}$ is the effective porosity, and V_{iv} and V_b are the volume of the interconnected void and the bulk volume, respectively.

Based on the gas expansion method, we built a measuring porosity platform on the autoclave in the laboratory, as shown in Figure 1. The principle of porosity measurement is Boyle's law. If the gas pressure was low enough and the temperature was not too low, then Boyle's law was approximately followed. This law is not dependent on the type of gas. In this apparatus, argon was used as test medium.

We defined the empty volume between needle valve 2 and needle valve 3 as the standard chamber, and between needle valve 2, needle valve 4, and the autoclave as the sample chamber. The connections between needle valves and other devices are through some capillary tubes with an inner diameter of 1/32 inch and an outer diameter of 1/16 inch. The empty volume of the standard chamber is composed of capillary pore volume, valve passage volume, and the inner void volume of pressure sensor. In addition, the empty volume of the sample chamber is composed of capillary pore volume, valve passage volume, the volume of autoclave plug hole, and sample pore volume. Owing to the small and irregular volume of the standard and sample chambers of this platform, the empty volumes of the two parts could not be measured by the direct method. Therefore, we used the system calibration method to avoid measuring the empty volume. The bulk volume of the rock was calculated by the volumetric formula, and its diameter and height were measured with a caliper.

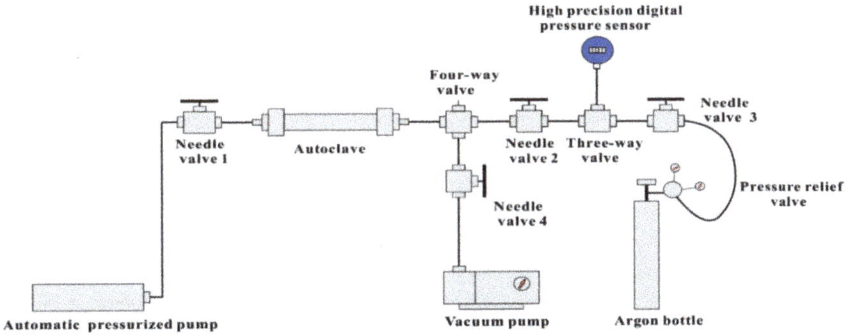

Figure 1. Schematic diagram of the rock porosity measurement apparatus.

Compared with previous devices, a vacuum pump was added, which was manufactured by Zhejiang Value Mechanical & Electrical Products Co., Ltd. (Taizhou, China), with an ultimate vacuum pressure of 2×10^{-1} Pa. Meanwhile, the automatic pressurized pump supported the confining pressure of the sample instead of the hydraulic oil pump, whose nominal pressure was 30,000 psi ($\approx$207 MPa). In addition, high-precision digital pressure sensor, manufactured by Xi'an Holder Instrument Co., Ltd. (Xi'an, China), with an accuracy class of $\pm$0.1% F-S., was a part of the platform.

3. Experimental Section

3.1. Sample Holder

As shown in Figure 2, the autoclave was used as the sample holder, which was crucial to the proposed porosity measurement apparatus. Due to the requirement for permeability measurements, on the side of the sample holder used by other investigators a port was opened for pressurizing confining pressure. The opened port would reduce its strength, therefore, the sample holder only withstood 90 MPa [31,32]. The proposed porosity measurement apparatus was only used for measuring porosity, thus, the side of the sample holder that did not open a port met our requirements. The apparatus consists of double metal thick-walled cylinders, the material of the inner cylinder is TC11 (tensile strength $\geq$1030 MPa, yield strength $\geq$900 MPa), the material of the outer cylinder is SUS 321 (tensile strength $\geq$550 MPa, yield strength $\geq$220 MPa). The safety factor was selected as 2.5, and the autoclave could withstand at least 400 MPa of pressure after a strength check.

The sample holder was mainly composed of an autoclave, two autoclave plugs with center hole, a fluorine rubber sleeve, a 304 stainless steel-ring, and metal/rock sample. The sample was wrapped with a fluorine rubber sleeve, and then connected to the end of the autoclave plug by the fluorine rubber sleeve. The connection between fluorine rubber sleeve and the end of the autoclave plug was tightened with a 304 stainless steel-ring. The sample holder could protect the sample from water immersion at a confining pressure of 200 MPa. The fluorine rubber sleeve was approximately 1 mm thick, and soft enough to conform to sample sides at low isostatic pressure [29]. The elastic modulus of the rubber sleeve was very small. Thus, in order to achieve a tight fit between the sample and the rubber sleeve, the initial isostatic pressure was set to 1 MPa.

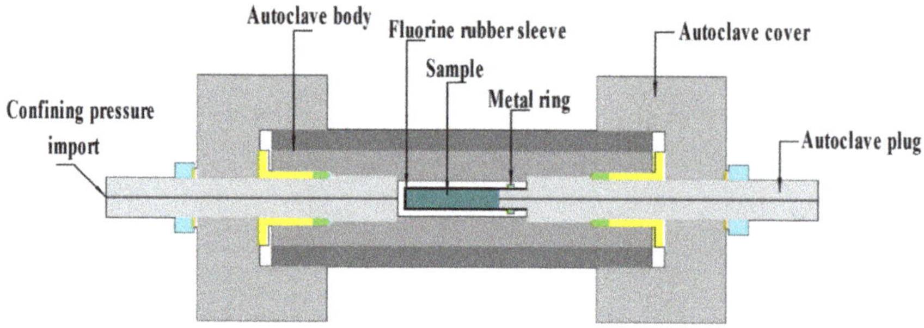

Figure 2. A sectional view of the autoclave.

3.2. Calibration of Equipment

A total of eleven 304 stainless steel cylindrical samples, each with a diameter of 10 mm and a height of 25 mm, were selected as the set of calibration samples. Different cylindrical samples contained different center-bored volumes, as shown in Figure 3a. The theoretical porosity of these samples was set to 0, 1%, 2%, 3%, 4%, 5%, 6%, 7%, 8%, 9%, and 10%. However, their practical porosity after processing is shown in Table 1.

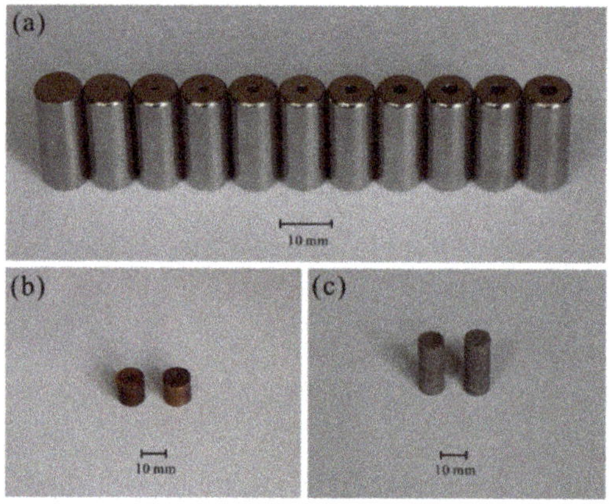

Figure 3. Photograph of the samples. (**a**) Cylindrical 304 stainless steel samples. (**b**) Cylindrical copper samples. (**c**) Cylindrical sandstone samples.

Table 1. The size and porosity of the processed calibration samples.

Sample Number	Center-Bore Diameter (mm)	Void Volume (mm^3)	Density (g/cm^3)	Effective Porosity (%)
S0	0	0	7.896	0
S1	0.93	17.002	7.862	0.87
S2	1.37	36.896	7.860	1.88
S3	1.69	56.145	7.851	2.87
S4	1.96	75.548	7.870	3.86
S5	2.20	95.144	7.865	4.86
S6	2.42	115.217	7.819	5.88
S7	2.64	137.008	7.864	7.00
S8	2.81	155.283	7.830	7.94
S9	2.96	172.166	7.826	8.78
S10	3.13	192.587	7.851	9.84

The porosity measurement apparatus was placed into a thermostatically controlled room, whose temperature was kept at 25 °C (±0.5 °C). The calibration of designed apparatus was performed as follows:

(1) One calibration sample was placed inside the fluorine rubber sleeve each time, assembling the entire platform.
(2) Needle valve 2 was closed and Needle valve 4 was opened at the same time. To make the fluorine rubber sleeve tightly fit the surface of sample, the sample chamber was vacuumed.
(3) To keep the sample under confining pressure of 1 MPa, pure water was pumped to pressurize the sample holder through an automatic pressure manual pump.
(4) To eliminate the vacuum of sample chamber, especially in sample, the sample chamber was injected with a certain amount of argon and kept for a while.
(5) To keep the initial pressure conditions consistent, argon, at a constant pressure of 180 kPa (±0.03 kPa), was pumped into the standard chamber.
(6) In order that the initial pressure of sample chamber was equal to atmospheric pressure, the top outlet of Four-way valve was opened.
(7) A few minutes later, the top outlet of Four-way valve was closed. For removing air from sample chamber, the sample chamber was vacuumed again. Meanwhile, the vacuuming time was set to 75 s so that the vacuum pressure of the sample chamber was consistent in each experiment.
(8) To flow argon into the sample chamber, Needle valve 4 was closed and then Needle valve 2 was opened. The value of equilibrium pressure was recorded when the gas pressure reached equilibrium states.
(9) In order to reduce the measurement error, the calibration experiment was repeated three times. The drop in gas pressure was calculated by averaging the three equilibrium pressures.

The graph between the drop in gas pressure and pore volume is shown in Figure 4. The fitting function $y = -0.00005x^2 + 0.0409x + 46.597$ was thus obtained.

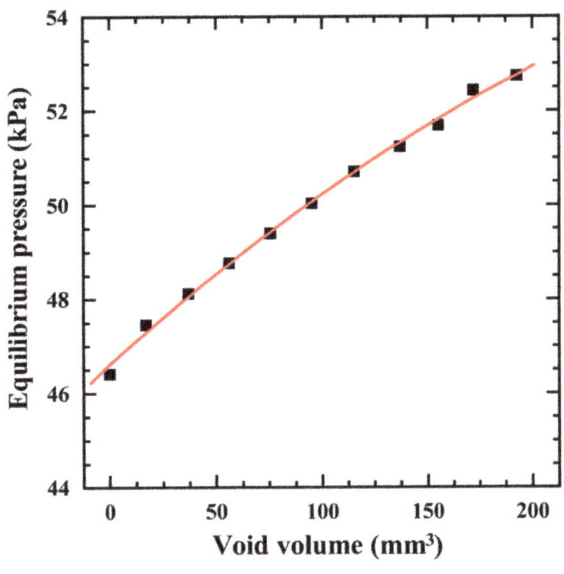

Figure 4. Calibration curve for the relationship between the drop in the gas pressure of the calibration sample and its void volume.

Under high isostatic pressure, the experimental confining pressure of our device went up to 200 MPa. We assumed that the elastic modulus of the rock grain was sufficiently large and that the change in its bulk volume could be ignored. Only the change in the volume of the rock void when the change in confining pressure was considered. Therefore, this fitting function is also applicable under high isostatic pressure.

3.3. *Testing of Equipment*

To verify the calibration, copper samples with voids of different sizes, as shown in Figure 3b, were used for testing. Their basic sizes and theoretical voids are shown in Table 2. The deviation in the test volume of the sample was less than 3.31 mm^3, and the corresponding porosity deviation was 0.41%. The porosity deviation was less than 0.5%, which is the maximal deviation that a common commercial measuring instrument for porosity is permitted.

Table 2. Basic dimensions and theoretical porosities of cylindrical copper samples.

Sample Number	Loss of Pressure Variation (kPa)	Void Volume (mm^3)	Bulk Volume (mm^3)	Determined Porosity (%)	Calculated Porosity (%)
Copper 01	49.56	81.40	778.70	10.32	10.45
Copper 02	50.97	129.78	805.83	15.69	16.10

4. Results and Discussion

As shown in Figure 3c, two cylindrical samples of sandstones were processed for the measurement experiment with the proposed apparatus. The basic dimensions of the sandstones are shown in Table 3.

Table 3. Basic dimensions and physical properties of sandstone.

Sample Number	Diameter (mm)	Height (mm)	Mass (g)	Bulk Density (g/cm^3)
Sandstone 1	10.48	26.35	5.50	2.23
Sandstone 2	10.44	28.08	5.99	2.49

The sandstones used in the laboratory were sourced from Lai Shui County in Hebei Province, China. The sandstones were cored and processed to an appropriate size, and oven-dried at 42 °C (±1 °C) for 20 h. Before porosity testing, the permeability of the sandstones was first tested by the pressure drop method, and these results are as shown in Figure 5. Although the permeability of the two sandstones exhibited some differences, permeability measurements indicated that effective porosity can be adequately measured by the designed apparatus. The isostatic pressure increased from 1 MPa to 200 MPa, and the porosity of the sandstones was then measured. as shown in Figure 6.

In general, hydrostatic volumetric compaction is exponentially related to stress, which also means that porosity is exponentially related to stress [33]. According to the fitted power function, the fitting functions of sandstone 1 and sandstone 2 were obtained as $y = 0.0541 \times x^{-0.326}$, $R^2 = 0.9891$, and $y = 0.0565 \times x^{-0.312}$, $R^2 = 0.9716$, respectively. To reach a porosity of 0.5%, 1.49 GPa of pressure was required for sandstone 1 and 2.37 GPa for sandstone 2. These results indicate obvious differences in the porosity of different sandstones, where this decreases markedly with increasing hydrostatic pressure. From a previous study, we know that narrow cracks are first closed in sandstone and that the closure pressure is very low. However, when the cavities of porous quartz sandstone achieve an aspect ratio of 1/10, its closure pressure increases significantly up to 10 GPa [34].

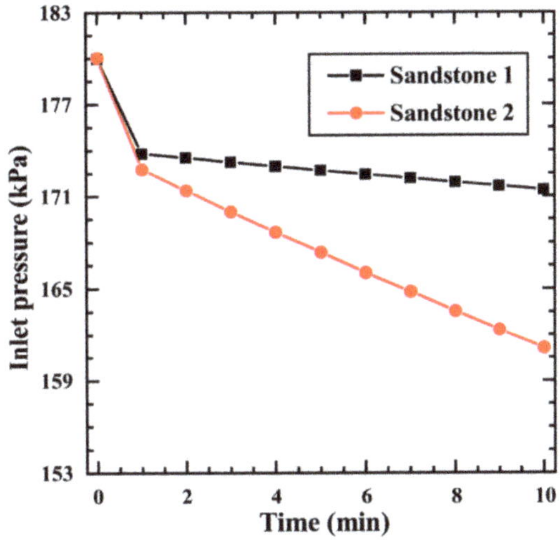

Figure 5. Relationship between inlet pressure and time.

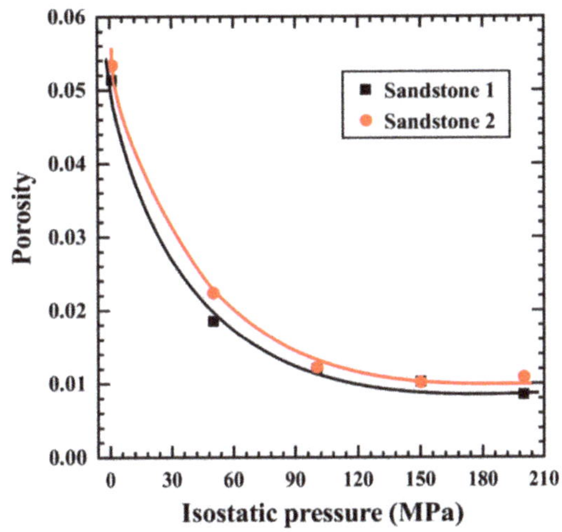

Figure 6. Relationship between porosity and isostatic pressures of two sandstones.

Some experimental parameters were not discussed in detail in a previous study [35], which easily caused systemic error. Therefore, these parameters deserved to be considered for the proposed apparatus, such as the time needed for vacuuming, choosing the initial gas pressure, and selection of the gas pressure equilibrium time.

4.1. Time Needed for Vacuuming

In each experiment, there are two vacuums. To verify sample tightness before calibrating the sample, the sample chamber was vacuumed in advance of the supply confining pressure, which could form self-sealing and reduce the clearance error between the sample and fluorine rubber sleeve. In addition, the sample chamber was vacuumed before measuring porosity, which could reduce the interference of air inside the sample chamber.

To reduce the error caused by vacuum gauges, we did not add it to the system in the actual measurements. The vacuum pressure was determined by controlling the time needed for vacuuming. The vacuum gauge was mounted on top of the sample chamber—

the error was ±0.5% F-S, and the sample chamber was vacuumed. The initial pressure was nearly 86.88 kPa at atmospheric pressure. After three repetitions of vacuuming, the vacuum pressure at different times was approximately the same, and was less than 1 kPa after more than 75 s. Meanwhile, the vacuum pressure can increase by approximately 0.52 kPa in 30 s or 0.73 kPa in 60 s. However, the operation time is usually less than 5 s; that is, the operation has little effect on the experimental results. Furthermore, the longer the time needed for vacuuming was, the smaller the vacuum degree could reach. To significantly improve measurement accuracy, we chose 75 s as the time needed for vacuuming.

4.2. Choosing Initial Gas Pressure

According to Boyle's law of gases, the lower the gas pressure is, the closer a gas is to an ideal gas. We used a high-precision digital pressure sensor with a measuring range of 0–200 kPa, consistent with the experimental requirements. To determine an appropriate initial pressure, 120 kPa,150 kPa, and 180 kPa were selected as candidate initial pressures to observe gas changes after pressure equilibrium had been reached. As shown in Figure 7, the linear fitting functions were, $y = 0.005x + 127.36$, $y = 0.008x + 107.14$, and $y = 0.027x + 87.67$, respectively. When the initial gas pressure was 180 kPa, the magnitude of the pressure change over time was small after equilibrium in gas pressure had been reached, that is, the slope was the smallest. We infer that this result was caused by ambient temperature interference. To reduce this interference, we chose 180 kPa as the initial gas pressure.

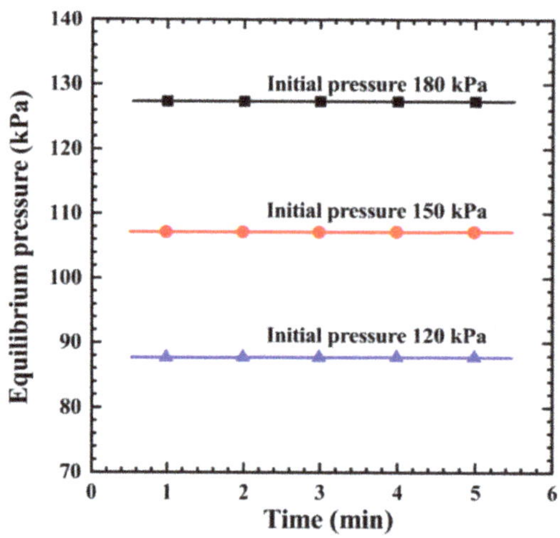

Figure 7. Relationship between gas equilibrium pressure and equilibrium time when the initial gas pressures were 120 kPa, 150 kPa, and 180 kPa.

4.3. Selection of the Gas Pressure Equilibrium Time

Compared to the metal sample, the rock sample including many micropores and pore canals that were extremely tortuous, thus, the time that gas pressure reached equilibrium was long in the rock sample porosity testing, which would also influence the measurement accuracy of rock porosity. Moreover, the gas pressure was also influenced by the ambient temperature. Therefore, the choice of gas pressure equilibrium time was important. To maximize the measurement accuracy of rock porosity, we selected the time within 1 min of when there was a drop of 0.2 kPa, and this was considered as the gas pressure equilibrium time. The gas pressure equilibrium time was generally more than 25 min unless the gas pressure stopped falling before then.

5. Conclusions

The rock porosity measuring apparatus that took the autoclave as the sample holder was built, which can achieve porosity determination under a high isostatic pressure of up to 200 MPa. The measurement error of the proposed apparatus after calibration achieved the same level as similar instruments. Furthermore, the porosity of sandstones under high isostatic pressures only had a small variation of less than 0.5%. Finally, a series of experiments were conducted to investigate the effects of experimental conditions on porosity measurements under different confining pressures. The main conclusions are as follows: (1) Increasing the initial gas pressure helps reduce the effect of ambient temperature on the porosity measurement results. (2) At the beginning of gas equilibrium, the pressure drop was quick but slowed down thereafter. In general, the gas pressure reached equilibrium in 25 min or sooner.

However, the temperature of the Earth's interior also influences the porosity of rocks, and should be considered in future work. Moreover, the other physical properties of rocks in the interior of the crust need to be examined.

Author Contributions: Conceptualization and methodology, H.L.; investigation, L.L.; preparation of experimental samples, L.L.; design and construction of experimental apparatus, H.L., L.L., H.Z., S.L. (Sen Lin) and S.L. (Shengbin Li); experimental analysis, H.L., L.L. and H.Z.; plotting, L.L.; writing—original draft preparation, L.L.; writing—review and editing, H.L., L.L., H.Z., S.L. (Sen Lin) and S.L. (Shengbin Li). All authors have read and agreed to the published version of the manuscript.

Funding: This study was supported by the Major State Research Development Program of China (Grant No. 2016YFC0601101), the National Natural Science Foundation of China (Grant No. 41827802), and the Science and Technology Foundation Project in Guizhou Province ([2019]1316).

Data Availability Statement: The data that support the findings of this study are available from the corresponding author upon reasonable request.

Acknowledgments: We thank Senior Engineer Shuangming Shan for providing the rock samples, and three anonymous reviewers and the editor for their helpful comments and suggestions.

Conflicts of Interest: The authors declare no conflict of interest.

References

1. Lebedev, E.; Ryzhenko, B.; Dorfman, A.; Zebrin, S.; Sokolova, N.; Burkhardt, H.; Morig, R.; Wulff, A. Influence of fluids on the elastic properties of sandstone at high pressure and temperature. *Geophys. Res. Lett.* **1996**, *23*, 3115–3118. [CrossRef]
2. Pan, L.; Jones, S.; Wang, X.; Guan, W.; Li, L. Re-evaluation of the porosity measurements under different confining pressures: A better appraisal of reservoir porosity. *AAPG Bull.* **2019**, *103*, 515–526. [CrossRef]
3. Liu, S.; Wang, H.; Xu, W.; Xiang, Z. The influence of water saturation and confining pressure to gas porosity and permeability of sandstone. *Environ. Earth Sci.* **2019**, *78*, 182. [CrossRef]
4. Beranek, L. Acoustic impedance of porous materials. *J. Acoust. Soc. Am.* **1942**, *13*, 248–260. [CrossRef]
5. Stumpf, F. Porosity-measurement apparatus. *Am. J. Phys.* **1960**, *28*, 503–504. [CrossRef]
6. Wyllie, M.; Gregory, A.; Gardner, L. Elastic wave velocities in heterogeneous and porous media. *Geophysics* **1956**, *21*, 41–70. [CrossRef]
7. Kassab, M.; Weller, A. Porosity estimation from compressional wave velocity: A study based on Egyptian sandstone formations. *J. Pet. Sci. Eng.* **2011**, *78*, 310–315. [CrossRef]
8. Domenico, S. Sandstone and limestone porosity determination from shear and compressional wave velocity. *Geophysics* **1984**, *49*, 637. [CrossRef]
9. Ge, Y.; Wu, W.; Wang, R.; Fu, X.; He, L. Thermal neutron cross section logging based on compensated neutron logging. *Appl. Radiat. Isot.* **2020**, *166*, 109317. [CrossRef]
10. Yu, H.; Sun, J.; Wang, J.; Gardner, R. Accuracy and borehole influences in pulsed neutron gamma density logging while drilling. *Appl. Radiat. Isot.* **2011**, *69*, 1313–1317. [CrossRef]
11. Mitchell, J.; Fordham, E. Contributed review: Nuclear magnetic resonance core analysis at 0.3 T. *Rev. Sci. Instrum.* **2014**, *85*, 111502. [CrossRef]
12. Withjack, E. Computed tomography for rock-property determination and fluid-flow visualization. *SPE Form. Eval.* **1988**, *3*, 696–704. [CrossRef]
13. Jin, J.; Kim, J.; Lee, J.; Oh, Y. Correlative multiple porosimetries for reservoir sandstones with adoption of a new reference-sample-guided computed-tomographic method. *Sci. Rep.* **2016**, *6*, 30250. [CrossRef]

14. Bodwadkar, S.; Reis, J. Porosity measurements of core samples using gamma-ray attenuation. *Nucl. Geophys.* **1994**, *8*, 61–78.
15. Iscan, A.; Kok, M. Porosity and permeability determinations in sandstone and limestone rocks using thin section analysis approach. *Energy Sources Part A-Recovery Util. Environ. Eff.* **2009**, *31*, 568–575. [CrossRef]
16. Vanbrakel, J.; Modry, S.; Svata, M. Mercury porosimetry—State of the art. *Powder Technol.* **1981**, *29*, 1–12. [CrossRef]
17. Salissou, Y.; Panneton, R. Pressure/mass method to measure open porosity of porous solids. *J. Appl. Phys.* **2007**, *101*, 142–143. [CrossRef]
18. Washburn, E.; Bunting, E. Determination of porosity by the method of gas expansion. *J. Am. Ceram. Soc.* **1922**, *5*, 112–129. [CrossRef]
19. Diamond, S. Mercury porosimetry: An inappropriate method for the measurement of pore size distributions in cement-based materials. *Cem. Concr. Res.* **2000**, *30*, 1517–1525. [CrossRef]
20. Leclaire, P.; Umnova, O.; Horoshenkov, K.; Maillet, L. Porosity measurement by comparison of air volumes. *Rev. Sci. Instrum.* **2003**, *74*, 1366–1370. [CrossRef]
21. Giesche, H. Mercury porosimetry: A general (practical) overview. *Part. Part. Syst. Charact.* **2006**, *23*, 9–19. [CrossRef]
22. Richmond, J.; Peterson, J.; Herschel, W. An improved volumenometer. *J. Am. Ceram. Soc.* **1943**, *26*, 127–131. [CrossRef]
23. Kampmeyer, P. The temperature dependence of viscosity for water and mercury. *J. Appl. Phys.* **1952**, *23*, 99–102. [CrossRef]
24. Rigby, M.; Smith, E. Viscosities of inert gases. *Trans. Faraday Soc.* **1966**, *62*, 54–58. [CrossRef]
25. Watanabe, N.; Ishibashi, T.; Tsuchiya, N.; Ohsaki, Y.; Tamagawa, T.; Tsuchiya, Y.; Okabe, H.; Ito, H. Geologic core holder with a CFR PEEK body for the X-ray CT-based numerical analysis of fracture flow under confining pressure. *Rock Mech. Rock Eng.* **2013**, *46*, 413–418. [CrossRef]
26. Shakerian, M.; Marica, F.; Afrough, A.; Goora, F.; Li, M.; Vashaee, S.; Balcom, B. A high-pressure metallic core holder for magnetic resonance based on Hastelloy-C. *Rev. Sci. Instrum.* **2017**, *88*, 123703. [CrossRef]
27. Wang, H.; Chen, Y.; You, L.; Cai, M. The development of multi-parameters measurement instrument uesd for high tempreature and high presure core. *J. Southwest Pet. Inst. Nat. Sci. Ed.* **2007**, *29*, 138–140.
28. Zhang, K.; Chen, Y.; Xu, H. High temperature and high pressure rock gas porosity automatic test system. *Petrol. Tub. Goods Instr.* **1995**, *9*, 79–85.
29. Keelan, D. Automated core measurement system for enhanced core data at overburden conditions. In Proceedings of the Society of Petroleum Engineers SPE Rocky Mountain Regional Meeting, Billings, MT, USA, 19–21 May 1986. [CrossRef]
30. Dotson, B.; Slobod, R.; Mccreery, P.; Spurlock, J. Porosity-measurement comparisons by five laboratories. *J. Pet. Technol.* **1951**, *3*, 341–346. [CrossRef]
31. Chen, X.; Caratini, G.; Davy, C.; Troadec, D.; Skoczylas, F. Coupled transport and poro-mechanical properties of a heat-treated mortar under confinement. *Cem. Concr. Res.* **2013**, *49*, 10–20. [CrossRef]
32. Pei, Y.; Agostini, F.; Skoczylas, F. The effects of high temperature heating on the gas permeability and porosity of a cementitious material. *Cem. Concr. Res.* **2017**, *95*, 141–151. [CrossRef]
33. Teeuw, D. Prediction of formation compaction from laboratory compressibility data. *Soc. Petrol. Eng. J.* **1971**, *11*, 263–271. [CrossRef]
34. Walsh, J. Effect of cracks on compressibility of rock. *J. Geophys. Res.* **1965**, *70*, 381–389. [CrossRef]
35. Tamari, S. Optimum design of the constant-volume gas pycnometer for determining the volume of solid particles. *Meas. Sci. Technol.* **2004**, *15*, 549–558. [CrossRef]

minerals

MDPI

Article

Effect of Different Mineralogical Proportions on the Electrical Conductivity of Dry Hot-Pressed Sintering Gabbro at High Temperatures and Pressures

Mengqi Wang [1,2], Lidong Dai [1,*], Haiying Hu [1,*], Wenqing Sun [1], Ziming Hu [1,2] and Chenxin Jing [1,2]

1 Key Laboratory of High-Temperature and High-Pressure Study of the Earth's Interior, Institute of Geochemistry, Chinese Academy of Sciences, Guiyang 550081, China; wangmengqi@vip.gyig.ac.cn (M.W.); sunwenqing@vip.gyig.ac.cn (W.S.); huziming@mail.gyig.ac.cn (Z.H.); jingchenxin@mail.gyig.ac.cn (C.J.)

2 University of Chinese Academy of Sciences, Beijing 100049, China

* Correspondence: dailidong@vip.gyig.ac.cn (L.D.); huhaiying@vip.gyig.ac.cn (H.H.)

Abstract: Electrical conductivities of the dry hot-pressed sintering gabbro with various mineralogical proportions (Cpx_XPl_{100-X}, X = 0, 10, 20, 30, 40, 50, 60, 70, 80, 90 and 100 vol% (the signals of Cpx and Pl denote clinopyroxene and plagioclase, respectively) were measured in the YJ-3000t multi-anvil pressure and Solartron-1260 impedance spectroscopy analyzer at temperatures of 773–1073 K and pressures of 1.0–3.0 GPa. At the given pressure conditions, the electrical conductivity and temperature conformed to an Arrhenius relation. For the fixed mineralogical composition of $Cpx_{50}Pl_{50}$, the electrical conductivities of the samples significantly increased with the rise of temperature, but slightly decreased with increasing pressure. Furthermore, the activation energy and activation volume were determined as 1.06 ± 0.12 eV and 6.00 ± 2.00 cm^3/mole, respectively. As for the various mineralogical compositions of dry gabbro, the electrical conductivities of the samples increased with the rise of volume percentage of clinopyroxene (*Cpx*) at 1.0 GPa. It is proposed that the main conduction mechanism is the small polaron, owing to the positive relation between the electrical conductivity and the iron content in samples. On the basis of these obtained conductivity results, laboratory-based electrical conductivity–depth profiles for the hot-pressed sintering gabbro with various mineralogical proportions and temperature gradients were successfully established. In conclusion, although the present acquired electrical conductivity results on the dry hot-pressed sintering gabbro with various mineralogical proportions cannot explain the high conductivity anomaly in the oceanic crust and West African craton, it can provide one reasonable constraint on the mineralogical composition in these representative gabbro-rich regions.

Keywords: electrical conductivity; gabbro; mineralogical proportions; high temperature; high pressure

Citation: Wang, M.; Dai, L.; Hu, H.; Sun, W.; Hu, Z.; Jing, C. Effect of Different Mineralogical Proportions on the Electrical Conductivity of Dry Hot-Pressed Sintering Gabbro at High Temperatures and Pressures. *Minerals* **2022**, *12*, 336. https://doi.org/10.3390/min12030336

Academic Editor: Julien Siebert

Received: 8 February 2022
Accepted: 4 March 2022
Published: 8 March 2022

1. Introduction

Results of the field geophysical observations from magnetotelluric (MT) and geomagnetic deep sounding (GDS) have confirmed that there is widespread existence of a large quantity of the anomalously high electrical conductivity phenomena in these typical regions of oceanic crust [1–3] and West African craton [4]. As a representative basic intrusive rock, it is well known that gabbro is widely outcropped in these global tectonic units of oceanic crust and craton regions, and its corresponding dominant mineralogical composition is constituted of clinopyroxene and plagioclase [5,6]. Therefore, a systematic investigation of the electrical conductivity of gabbro at high temperature and high pressure is crucial to deeply explore the cause of high conductivity anomaly in these regions.

As for natural gabbro, there are a large quantity of previously available reported electrical conductivity results using the piston-cylinder and multi-anvil press at conditions of high temperature and high pressure [7–10]. Wang et al. [7] measured the electrical conductivities of dry natural gabbro at pressures of 1.0–2.0 GPa and temperatures of

593–1173 K and found that the electrical conductivity significantly decreased by two orders of magnitude with the rise of pressure. The electrical conductivity and P-wave elastic velocity for natural biotite-bearing gabbro were determined by Bai et al. [8] within the temperature ranges from 293 K to 1373 K and pressure ranges from 1.0 GPa to 2.0 GPa. They proposed that the dehydration reaction of the accessory of hydrous biotite occurred at the critical temperature ranges of 1073–1123 K, and thus resulted in the appearance of the partial melting. Dai et al. [9] conducted electrical conductivity on the natural hydrous gabbro at conditions of 623–1173 K and 0.5–2.0 GPa, and three different solid oxygen buffers to control the oxygen fugacity (e.g., Cu + CuO, Ni + NiO, and Mo + MoO_2). Their mineralogical compositions of natural gabbro were composed of 60% plagioclase, 39% augite, and ~1% minor mineral phases. Some hydrogen-related defects in the natural hydrous gabbro with the relatively lower concentration were considered as the main charge carrier because a negative dependence relationship between the electrical conductivity and oxygen fugacity was observed by Dai et al. [9]. More recently, Saito and Bagdassarov [10] performed laboratory measurements of the electrical conductivity for gabbro samples in the Oman ophiolite at 523–1181 K and 0.6–0.8 GPa using the piston-cylinder high-pressure apparatus. In comprehensive considerations of all previously acquired conductivity results, we find that there exists one obvious discrepancy of almost three magnitude orders in the electrical conductivity of natural gabbro, which possibly originates from the differentiation of mineralogical proportion, water content, and accessory minerals.

As usual, the laboratory-based electrical conductivity of rock is highly correlated with its main constituent minerals at high temperate and high pressure. As two dominant constituent minerals in gabbro, electrical conductivities of clinopyroxene and plagioclase single crystals have been studied extensively. Yang et al. [11] obtained the electrical conductivities of lower crustal clinopyroxene at temperatures of 523–1273 K and pressures of 0.6–1.2 GPa in a piston-cylinder press. Electrical conductivities of anhydrous and hydrous clinopyroxene were measured by Zhao and Yoshino [12] at temperatures of 600–1200 K and 1.5 GPa in a DIA-type multi-anvil apparatus. By virtue of the Wayne Kerr B642 alternating current analyzer at a fixed frequency of 1.6 kHz, electrical conductivities of natural and synthetic polycrystalline diopside were measured at temperatures of 913–1337 K and pressure of 1.0 GPa [13]. Liu et al. [14] carried out the impedance spectroscopic measurements on omphacite single crystals containing 7.92 wt% iron and 0.0085 wt% water at temperatures of 623–973 K and 1.0 GPa in a piston-cylinder press. Sun et al. [15] researched the electrical conductivities of the clinopyroxene–NaCl–H_2O systems with various fluid fractions and salinities at 673–973 K and 1.0 GPa, and they found that the NaCl-bearing aqueous fluids in clinopyroxene can greatly enhance the electrical conductivity of fluid-bearing sample. Except for these abovementioned electrical conductivity results on clinopyroxene, Yang et al. [16] also performed the measurement of the electrical conductivity of lower crustal plagioclase (its corresponding chemical composition is $An_{66}Ab_{32}Or_2$. Here, *An*, *Ab*, and *Or* denote anorthite, albite, and orthoclase, respectively) at temperatures of 573–1273 K and pressures of 0.6–1.2 GPa, and further, the correlation between the electrical conductivity of plagioclase and water content was successfully established. It is well known that plagioclase is regarded as the solid solution of albite and anorthite. To explore the electrical transport mechanism for the feldspar family, Hu et al. [17,18] measured electrical conductivities of albite and anorthite within the temperature ranges from 773 K to 1073 K and pressure ranges from 1.0 GPa to 3.0 GPa in a YJ-3000t multi-anvil press. They found that the dominant conduction mechanisms for albite and anorthite are of the alkali ion, i.e., the monovalent cationic sodium and bivalent cationic calcium, respectively.

Because of the existence of heterogeneity in the mineralogical compositions and its corresponding mineralogical proportion between the clinopyroxene and plagioclase in different gabbro samples, it is impossible to directly extrapolate the electrical conductivity of gabbro from its main constituent minerals. Therefore, it is crucial to make a systematic study of electrical conductivity for gabbro with different mineralogical proportions at certain mineralogical compositions at high temperature and high pressure.

In the present study, we researched the electrical conductivities of dry hot-pressed sintering gabbro with various mineralogical proportions (Cpx_XPl_{100-X}, X = 0, 10, 20, 30, 40, 50, 60, 70, 80, 90 and 100 vol% at temperatures of 773–1073 K and pressures of 1.0–3.0 GPa. Further, the functional relationship between the electrical conductivity of sample and influential factors (e.g., temperature, pressure, and mineralogical proportion) was deeply explored. Based on these obtained electrical conductivity results and various thermodynamic parameters, we constructed the laboratory-based electrical conductivity–depth profile. Finally, the anomalously high electrical conductivity phenomena in oceanic crust and West African craton are discussed in detail.

2. Experimental Procedures

2.1. Sample Preparation

The natural gem-grade clinopyroxene (*Cpx*) and plagioclase (*Pl*) single crystals were chosen as the starting materials in order to synthesize the gabbro. The results of chemical composition of the main constituent minerals in the hot-pressed sintering gabbro was analyzed by virtue of the electron microprobe, as shown in Table 1.

Table 1. The chemical composition of the main constituent minerals in the hot-pressed sintering gabbro by virtue of the electron microprobe analysis.

Oxides (wt%)	Clinopyroxene (*Cpx*)	Plagioclase (*Pl*)
SiO_2	53.90	52.94
TiO_2	0.13	0.03
Al_2O_3	0.25	26.72
FeO	1.62	0.85
MnO	0.06	0.01
MgO	17.90	0.03
CaO	24.71	13.22
Na_2O	0.22	5.39
K_2O	0.00	0.06
NiO	0.02	0.00
Cr_2O_3	0.23	0.00
Total	99.04	99.25

The *Cpx* and *Pl* single crystal grains were cleaned in a mixture of distilled water, acetone, and ethanol, and then crushed and ground to powders with the grain size of 25 μm. Then, the mineral powders were homogeneously mixed with various volume percentages (Cpx_xPl_{100-x}, X = 0, 10, 20, 30, 40, 50, 60, 70, 80, 90 and 100 vol%) in the agate mortar and loaded into a copper–nickel doubly encapsulated sample chamber. The copper tube has a diameter of 8.5 mm, a thickness of 0.5 mm, and a height of 18.0 mm. To avoid interdiffusion between the copper tube and sample powder, 0.025 mm of nickel foil was adopted to encapsulate it. Subsequently, the sample assemblage was installed in the YJ-3000t multi-anvil high-pressure apparatus in order to complete the hot-pressed sintering of each gabbro with various mineralogical proportions. At the constant pressure of 1.0 GPa, the temperature in the sample chamber was 1073 K for 4 h. After that, the recovered samples were cut and polished into cylinders with diameters of 6.0 mm and heights of 4.0 mm. Lastly, the hot-pressed sintering gabbro was kept in the vacuum dry furnace at 423 K for at least 12 h before the in situ electrical conductivity measurements.

To identify the distribution of the minerals in hot-pressed sintering gabbro samples, four representative backscattered electron images of dry hot-pressed sintering gabbro with various mineralogical proportions were observed by using scanning electron microscope, as displayed in detail in Figure 1. As a whole, the distributions of *Cpx* and *Pl* grains in the acquired synthetic gabbro are almost homogeneous, and the magnitude of grain size is ~25 μm. As pointed out by Yang and Heidelbach [19], when the grain size of hot-pressed sintering synthetic clinopyroxene is larger than 10 μm, the influence of grain size on the electrical conductivity of a sample can be negligible under conditions of similar temperature and pressure ranges. Thus, the effect of grain size on the electrical conductivity of synthetic gabbro did not need to be considered in this study.

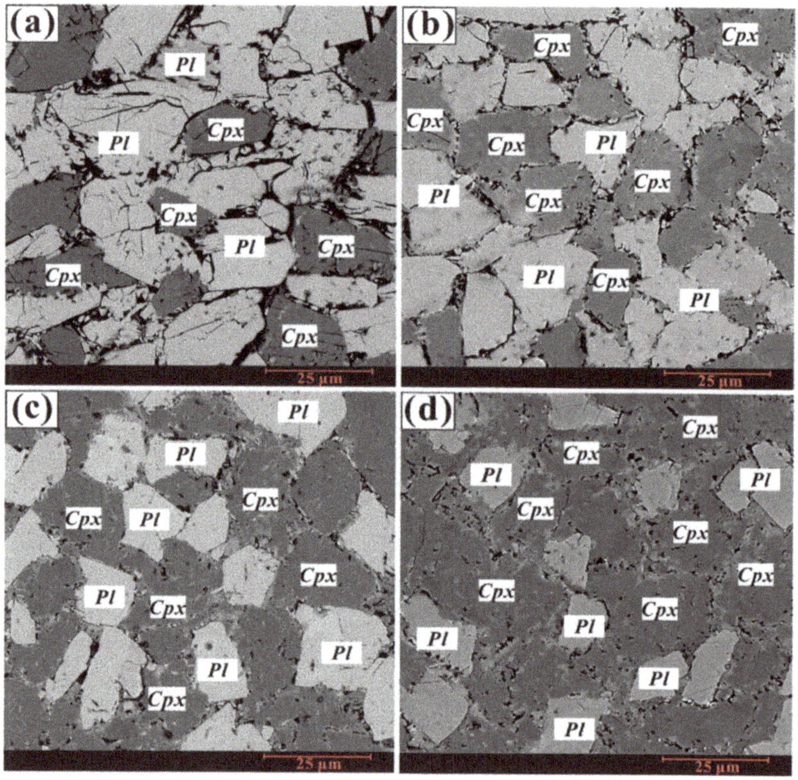

Figure 1. Four representative backscattered electron images of dry hot-pressed sintering gabbro by using scanning electron microscope with different mineralogical proportions: (**a**) $Cpx_{20}Pl_{80}$, (**b**) $Cpx_{40}Pl_{60}$, (**c**) $Cpx_{60}Pl_{40}$, and (**d**) $Cpx_{80}Pl_{20}$. *Cpx* and *Pl* represent clinopyroxene and plagioclase, respectively.

To accurately determine the water contents in the hot-pressed sintering gabbro before and after electrical conductivity experiments, the infrared spectra of samples were measured for each individual constituent *Cpx* and *Pl* mineral phases by a Fourier transform infrared spectrometer (Vertex-70V and Hyperion-1000 infrared microscope). The samples for Fourier transform infrared spectroscopy analysis (FT-IR) were double-polished up to a thickness of less than 200 μm. As shown in Figure 2, the FT-IR absorption spectra for clinopyroxene and plagioclase were conducted within the wavenumber range of 3000–4000 cm^{-1} in the hot-pressed sintering gabbro, respectively. By virtue of the representative hydrogen-related bands within the range of 3000–3750 cm^{-1}, the water content in the sample was calculated by the modified equation of the Beer–Lambert law [20]. The obtained water content in the sample is less than 25 ppm wt, which is almost anhydrous. A detailed description for the FT-IR measurement was given in the previous reference [21].

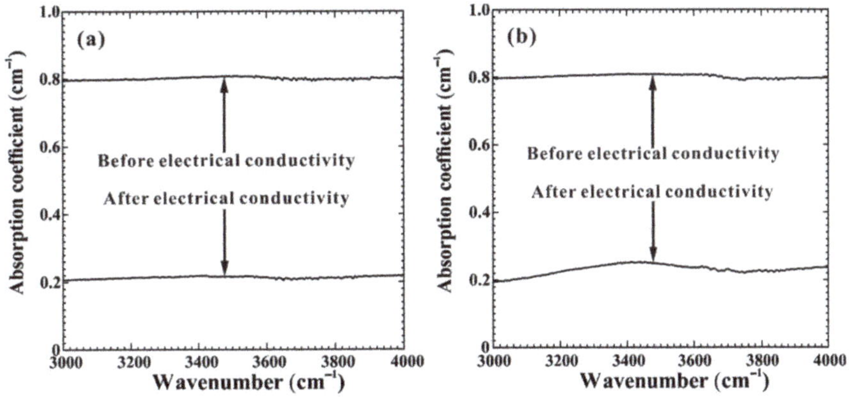

Figure 2. FT-IR spectra in the dry hot-pressed sintering gabbro before and after electrical conductivity measurements within the wavenumber range of 3000–4000 cm^{-1}. Here, (**a**,**b**) represent the clinopyroxene and plagioclase in the dry hot-pressed sintering gabbro, respectively.

2.2. High-Pressure Cell and Impedance Measurements

Electrical conductivity measurements were performed with the YJ-3000t multi-anvil apparatus and a Solartron-1260 Impedance/Gain-phase analyzer. Detailed measurement principles and experimental procedures are described by Dai et al. [22] and Hu et al. [23]. Figure 3 shows the schematic diagram of the experimental assemblage for electrical conductivity measurements at high pressure and temperature. Pyrophyllite with edge length of 32.5 mm was used as the pressure medium. The heater with a thickness of 0.5 mm consisted of three-layer stainless-steel sheets. The alumina and magnesia sleeves were adopted to provide a good insulation environment. The nickel foil (thickness: 0.025 mm) between the insulation sleeves was installed, which was directly linked to the Earth line to shield against the spurious signal interference and external electromagnetism. All these parts were baked at 1073 K for 5 h in a muffle furnace before the electrical conductivity of sample assembly. Two symmetric nickel disks with thickness of 0.5 mm and correspondent diameter of 6.0 mm were installed at both ends of the sample as measurement electrodes. A K-type thermocouple was used to monitor the temperature during the process of electrical conductivity measurements at high pressure.

For each electrical conductivity measurement, pressure was enhanced with a rate of 1.0 GPa/h to the desired value. Under the constant pressure condition, the temperature was gradually increased with a speed of 120 K/h. At a given temperature and pressure, the impedance spectra of samples were measured on the Solartron-1260 Impedance/Gain-phase analyzer with an applied voltage of 1.0 V signal voltage within the frequency range of 10^{-1}–10^{6} Hz. To check the reproducibility of electrical conductivity result, each electrical conductivity measurement in the sample was continuously measured at least two heating and cooling cycles. The uncertainties from the experimental pressure and temperature were less than 0.1 GPa and 5 K, respectively.

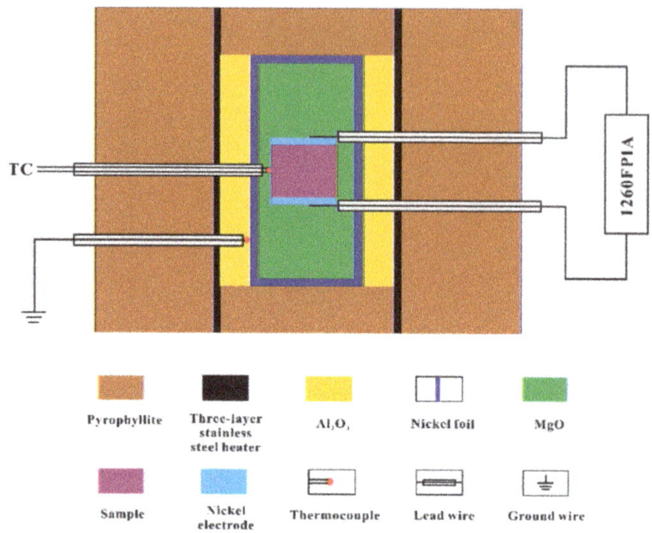

Figure 3. The experimental setup for electrical conductivity measurements at high temperature and high pressure.

3. Results

In the experiments, we measured the electrical conductivities of dry gabbro with various mineralogical proportions (Cpx_xPl_{100-x}, X = 0, 10, 20, 30, 40, 50, 60, 70, 80, 90 and 100 vol%) at temperature ranges of 773–1073 K and 1.0 GPa. To explore the effect of pressure on the conductivity of the dry gabbro, the electrical conductivities of $Cpx_{50}Pl_{50}$ sample were measured at temperature ranges of 773–1073 K and pressure ranges of 1.0–3.0 GPa.

Typical complex impedance spectra of hot-pressed sintering gabbro ($Cpx_{50}Pl_{50}$) at temperature ranges of 773–1073 K and pressure of 1.0 GPa are shown in Figure 4. The impedance spectra arcs were composed of an ideal semicircular arc within the high-frequency range from 10^6 Hz to ~1 Hz and another extra small tail at the frequency range from ~1 Hz to 10^{-1} Hz, which reflected the electrical transport process of grain interior and polarization effect at the sample–electrode interface, respectively [24]. Therefore, the equivalent circuit composed of the series connection of R_S-CPE_S (R_S and CPE_S represent the resistance and the constant phase element from the grain interior, respectively) and R_E-CPE_E (R_E and CPE_E represent the resistance and the constant phase element from the polarization effect at the sample–electrode interface, respectively) were adopted to fit the impedance spectra arc and acquire the conductivity results.

The electrical conductivities of the samples were calculated by the following equation,

$$\sigma = L \ / \ SR \tag{1}$$

where σ is the electrical conductivity of the sample (S/m), R is the resistance (Ω), L is the length of the sample (m), and S is the cross-sectional area of the electrode (m^2). The electrical conductivities of the hot-pressed sintering gabbro and the temperatures conformed to the Arrhenius relation,

$$\sigma = \sigma_0 \exp(-\Delta H \ / kT) \tag{2}$$

where σ_0 is the pre-exponential factor (S/m), k is the Boltzmann constant, T is the absolute temperature (K), and ΔH is the activation enthalpy (eV). The fitting parameters for the electrical conductivities are listed in Table 2.

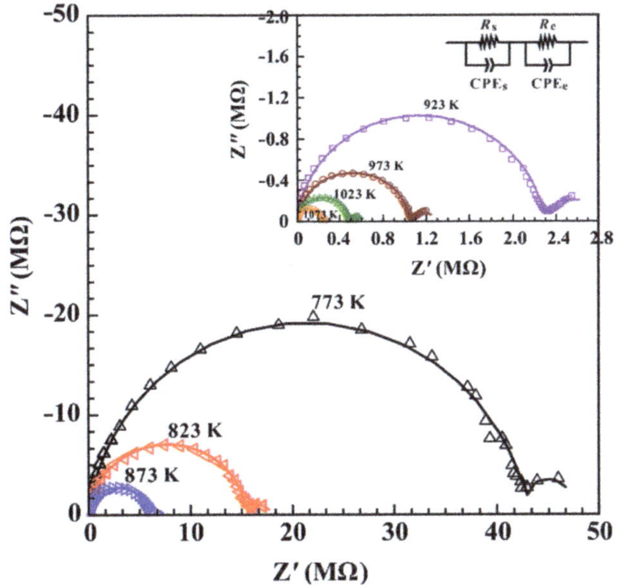

Figure 4. Typical complex impedance spectra for the $Cpx_{50}Pl_{50}$ sample, obtained at temperatures of 773–1073 K and pressure of 1.0 GPa in the frequency range from 10^{-1} Hz to 10^{6} Hz. Z' and Z'' are the real and imaginary parts of the complex impedance, respectively. The equivalent circuit that is composed of the series connection of R_S-CPE_S and R_E-CPE_E (R_S and CPE_S represent the resistance and the constant-phase element of the bulk impedance for the sample, respectively; R_E and CPE_E represent the resistance and the constant phase element from the polarization effect of the sample-electrode interface, respectively) is selected to model the impedance semicircles. The fitting results for the experimental data are displayed by using the solid line.

Table 2. Fitted parameters of the Arrhenius relation for the electrical conductivity of dry hot-pressed sintering gabbro samples. P is pressure, T is temperature, σ_0 is the pre-exponential factor, ΔH is activation enthalpy, and R^2 is the linear correction.

Run No.	Sample	P (GPa)	T (K)	Log [σ_0 (S/m)]	ΔH (eV)	R^2
DW02	Dry Pl	1	773–1073	2.43 ± 0.16	1.28 ± 0.03	99.68
DW04	$Cpx_{10}Pl_{90}$	1	773–1073	2.36 ± 0.07	1.26 ± 0.01	99.94
DW06	$Cpx_{20}Pl_{80}$	1	773–1073	2.36 ± 0.09	1.25 ± 0.02	99.90
DW08	$Cpx_{30}Pl_{70}$	1	773–1073	2.38 ± 0.12	1.24 ± 0.02	99.81
DW10	$Cpx_{40}Pl_{60}$	1	773–1073	2.40 ± 0.07	1.23 ± 0.01	99.93
DW12	$Cpx_{50}Pl_{50}$	1	773–1073	2.46 ± 0.06	1.23 ± 0.01	99.91
DW26	$Cpx_{50}Pl_{50}$	2	773–1073	2.77 ± 0.21	1.32 ± 0.04	99.52
DW28	$Cpx_{50}Pl_{50}$	3	773–1073	2.88 ± 0.23	1.37 ± 0.05	99.33
DW14	$Cpx_{60}Pl_{40}$	1	773–1073	2.52 ± 0.08	1.23 ± 0.01	99.78
DW16	$Cpx_{70}Pl_{30}$	1	773–1073	2.56 ± 0.11	1.22 ± 0.01	99.96
DW18	$Cpx_{80}Pl_{20}$	1	773–1073	2.53 ± 0.06	1.21 ± 0.01	99.95
DW20	$Cpx_{90}Pl_{10}$	1	773–1073	2.31 ± 0.07	1.15 ± 0.01	99.93
DW22	Dry Cpx	1	773–1073	2.02 ± 0.01	1.09 ± 0.01	99.95

For two fixed mineralogical proportions of $Cpx_{30}Pl_{70}$ and $Cpx_{70}Pl_{30}$ gabbro, the electrical conductivities of samples among two heating–cooling cycles at a pressure of 1.0 GPa are shown in Figure 5, respectively. In the first heating process, it is clear that a relatively higher value of electrical conductivity in gabbro was observed, which is possibly owing to the disequilibrium of thermal transference at 1.0 GPa. Subsequently, the electrical conductivity of the sample is almost reproducible in the first cooling and another heating–cooling cycle.

A good reproducibility in the electrical conductivity of sample denotes a completely stable and equivalent state to be reached in the sample chamber during the process of electrical conductivity measurement, which will be also adopted to fit the following electrical conductivity results.

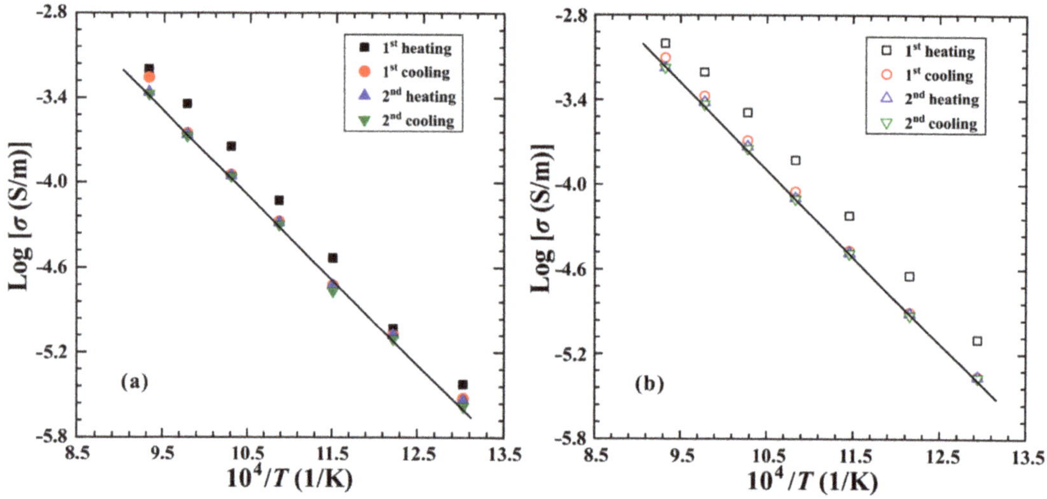

Figure 5. The electrical conductivities of the hot-pressed sintering gabbro among two heating–cooling cycles at a pressure of 1.0 GPa. Here, (**a**,**b**) represent the samples of $Cpx_{30}Pl_{70}$ and $Cpx_{70}Pl_{30}$, respectively.

4. Discussions

4.1. Influence of Pressure on Electrical Conductivity

To identify the effect of pressure on the electrical conductivity of gabbro, we measured the electrical conductivities of dry hot-pressed sintering gabbro ($Cpx_{50}Pl_{50}$) at pressures of 1.0–3.0 GPa. As shown in Figure 6, the electrical conductivity of the sample slightly decreases with the rise of pressure at temperature ranges of 773–1073 K. Further, the pre-exponential factor increased from 2.51×10^2 S/m to 7.59×10^2 S/m and activation enthalpy raised from 1.23 eV to 1.35 eV with the rise of 1.0 GPa to 3.0 GPa, respectively. In the present studies, the influence of pressure on the electrical conductivity can be described as

$$\sigma = A_0 \cdot (1 - BP) \cdot \exp(\frac{\Delta U + P\Delta V}{kT}) \quad (3)$$

Here, the pre-exponential factor (σ_0) and activation enthalpy (ΔH) of pressure dependence can be described as the equations of $\sigma_0 = A_0(1 - BP)$ and $\Delta H = \Delta U + P\Delta V$, where ΔU is the activation energy (eV), ΔV is the activation volume (cm^3/mole), P is pressure (GPa), and B is a constant with a dimension of 1/GPa. Therefore, the globally fitting data was adopted to fit it, and the fitting results are shown in Table 3. The logarithmic electrical conductivity of hot-pressed sintering sample and the inverse temperature follows a good linear relation, which reveals only one conduction mechanism operating the electrical transport within our experimental pressure ranges. By virtue of the Arrhenius relationship between the electrical conductivity and temperature at different pressures, we can extrapolate a linear relationship between log σ and the reciprocal temperature at atmospheric pressure. Furthermore, the pre-exponential factor and activation enthalpy at room pressure are calculated as 1.99×10^2 S/m and 1.17 eV, respectively. According to Equation (3) and Table 3, the activation energy and activation volume of the dry gabbro ($Cpx_{50}Pl_{50}$) can be determined as 1.06 ± 0.12 eV and 6.00 ± 2.00 cm^3/mole.

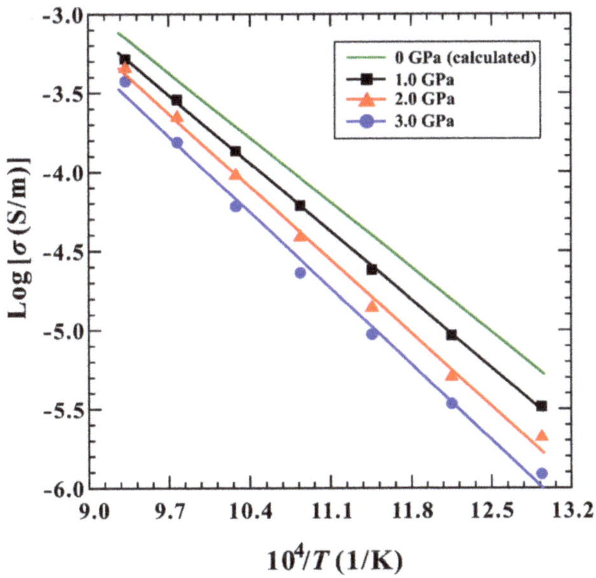

Figure 6. Influence of pressure on the electrical conductivity of dry hot-pressed sintering gabbro ($Cpx_{50}Pl_{50}$) at the temperature ranges of 773–1073 K.

Table 3. Parameter values for the electrical conductivity of dry hot-pressed sintering gabbro $Cpx_{50}Pl_{50}$ (units: σ_0: S/m, ΔU: kJ/mole, ΔV: cm^3/mole, P: GPa). The equation $\sigma = \sigma_0 \exp(\frac{\Delta U + P\Delta V}{kT})$ is adopted for the globally fitting of electrical conductivity data in sample. Considering the strong dependence on experimental data of σ_0 value on pressure, we used the relation $\sigma_0 = A_0(1 - \mathrm{BP})$ (where the unit for B is 1/GPa).

σ_0	ΔU (kJ/Mole)	ΔV (cm^3/Mole)
A_0 = 92.40 ± 14 (S/m) B = 0.23 ± 0.02 (1/GPa)	102 ± 12	6.00 ± 2.00

In the present study, we observed a negative dependence relation for the electrical conductivity of dry hot-pressed sintering gabbro ($Cpx_{50}Pl_{50}$) on the pressure, which is in good consistency with previously reported results [8]. In addition, Dai et al. [9] also investigated the electrical conductivity of natural gabbro at conditions of temperature ranges from 623 K to 1173 K, pressure ranges from 0.5 GPa to 2.0 GPa, and controlled different oxygen fugacities by three solid buffers (Cu + CuO, Ni + NiO, and Mo + MoO_2). On the contrary, a positive dependence relationship between the effect of electrical conductivity of natural gabbro and pressure was observed at a similar experimental condition. It is possibly related to the water content of the gabbro sample, which is also found to have a similar pressure effect to those of previously available electrical conductivity of anhydrous pyrope-rich and almandine-rich garnet single crystals at high temperature and high pressure [9,25,26].

4.2. *Influence of Mineralogical Proportions on Electrical Conductivity*

We have already measured the electrical conductivity on the hot-pressed sintering gabbro with various mineralogical proportions (Cpx_XPl_{100-X}, X = 0, 10, 20, 30, 40, 50, 60, 70, 80, 90 and 100 vol%) under conditions of 873–1473 K, 1.0 GPa, and the Ni–NiO solid buffer. The chemical composition of gabbro sample is well controlled by changing the mineralogical proportion of clinopyroxene and plagioclase. At a fixed pressure of

1.0 GPa, the effects of mineralogical proportions on the electrical conductivity of gabbro are illustrated in Figures 7 and 8.

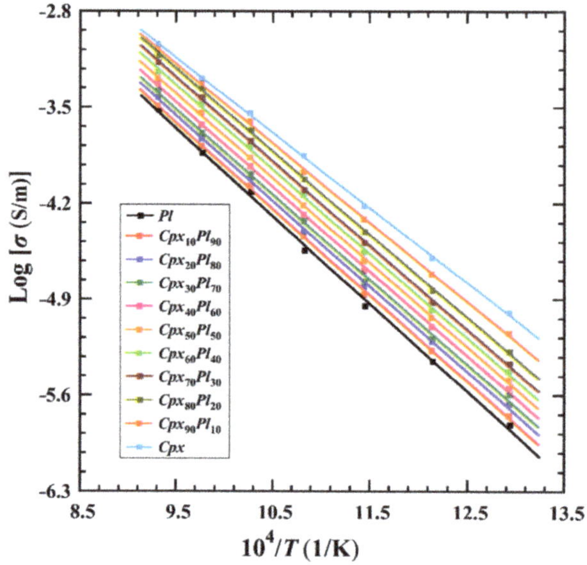

Figure 7. Logarithmic electrical conductivities of dry hot-pressed sintering gabbro with different mineralogical proportions versus reciprocal temperature at temperature ranges of 773–1073 K and 1.0 GPa.

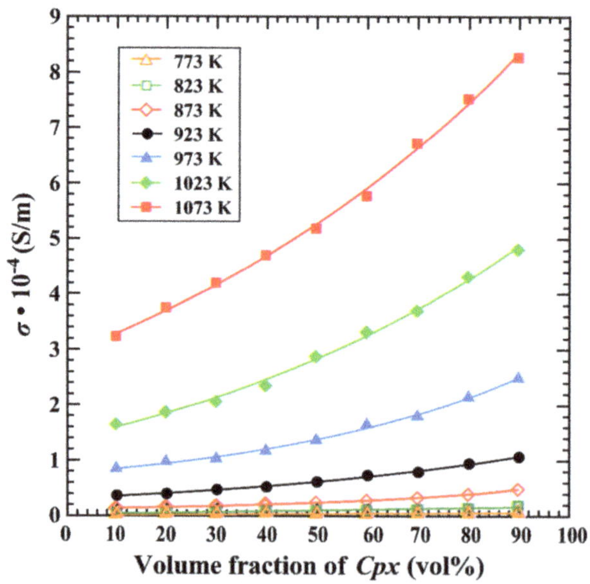

Figure 8. Effect of clinopyroxene content on the electrical conductivity of dry hot-pressed sintering gabbro samples with various mineralogical proportions at temperatures of 773–1073 K and 1.0 GPa.

Figure 7 shows the dependence of the electrical conductivity on temperature for gabbro samples with different mineralogical proportions. With the rise of temperature, the electrical conductivity of each hot-pressed sintering gabbro gradually increases accordingly.

For each correspondent mineralogical proportion (Cpx_XPl_{100-X}, X = 0, 10, 20, 30, 40, 50, 60, 70, 80, 90 and 100 vol%), the logarithm of electrical conductivity of the sample and reciprocal temperature follows a good linear relation, which reveals only one individual conduction mechanism operating the electrical transport process at high pressures.

As for each temperature point, a detailed dependence of the electrical conductivity of hot-pressed sintering gabbro on the volume percentage of clinopyroxene was illustrated in Figure 8. It is clear that the electrical conductivity of hot-pressed sintering gabbro shows a strong dependence on the variation of the volume percentage of clinopyroxene, i.e., electrical conductivity of the sample increases with the rise of the volume percentage of clinopyroxene at a constant temperature condition. At the same mineralogical proportion, the electrical conductivity of dry gabbro gradually increases with the rise of the temperature. Noteworthily, the dependence relation of electrical conductivity of gabbro on the variation of volume percentage of clinopyroxene tends to strengthen when the measured temperature is enhanced from 773 K to 1073 K. As one example at the fixed temperature of 1073 K, when the volume percentage of clinopyroxene is increased from 10% to 90%, the electrical conductivity of hot-pressed sintering gabbro is enhanced from about 3.23 × 10^{-4} S/m to 8.27 × 10^{-4} S/m accordingly.

In order to further explore the mineralogical proportion on the electrical conductivity of the hot-pressed sintering gabbro, a function relation for the electrical conductivity of gabbro on the volume percentage of clinopyroxene was established as follows,

$$\sigma = CX_{Cpx}^{\alpha}\exp(-\frac{D+\beta X_{\mathrm{Cpx}}^{\gamma}}{kT}) \tag{4}$$

In here, the pre-exponential factor (σ_0) and activation enthalpy (ΔH) of pressure dependence can be described as the equations of $\sigma_0 = A_0(1-BP)$ and $\Delta H = D + \beta X_{Cpx}^{\gamma}$, where the C is the pre-exponential factor, the D stands for the activation enthalpy at very small values of X_{Cpx}, X_{Cpx} stands for the volume percentage of clinopyroxene, and all of these parameters including α, β and γ are constants. Therefore, the globally fitting data was adopted to fit it, and the fitting results are shown in Table 4.

Table 4. Fitting parameters of activation enthalpy and pre-exponential factor for the electrical conductivity of dry hot-pressed sintering gabbro with various mineralogical proportions at temperatures of 773–1073 K and 1.0 GPa obtained from $\sigma = CX_{Cpx}^{\alpha}\exp(-\frac{D+\beta X_{Cpx}^{\gamma}}{kT})$. In here, the C stands for the pre-exponential factor, the D stands for the activation enthalpy at very small values of X_{Cpx}, X_{Cpx} stands for the volume percentage of clinopyroxene, and all of these parameters including α, β and γ are constants.

X_{Cpx}	log C	D	α	β	γ
$0.1 \leq X_{Cpx} \leq 0.9$	2.47 ± 1.34	1.25 ± 0.01	0.12 ± 0.05	−8.31 ± 0.91	1.60 ± 0.24

5. Comparisons with Previous Studies

As the dominant endmember of constituent mineral in gabbro, the mineralogical content of clinopyroxene has a significant influence on the electrical conductivity of the sample at conditions of 773–1073 K, 1.0 GPa and controlled oxygen fugacity by a Ni–NiO solid buffer. As displayed in Figure 9a, five previously reported results on the electrical conductivity of natural clinopyroxene were employed to compare it in detail [11–15]. Previous results have already confirmed that the iron content and water content are two crucial influential ingredients in the electrical conductivity of Fe-bearing silicate minerals at high temperatures and high pressures [25,27–30]. In the present studies, some microscopic observations from the electron microprobe and the Fourier transform infrared spectroscopy measurements confirmed that the iron content of clinopyroxene is ~1.62 wt% and belonging to a typically anhydrous sample. The electrical conductivities for our dry clinopyroxene are enhanced from 1.01 × 10^{-5} S/m to 9.19 × 10^{-4} S/m as the temperature increases from

773 K to 1073 K at 1.0 GPa. We note that, under the similar experimental conditions, our present results on the electrical conductivity of dry clinopyroxene are in good agreement with the anhydrous clinopyroxene reported by Yang et al. [11] and Zhao and Yoshino [12], but about 1.5 orders of magnitude lower than the electrical conductivities of hydrous sample from Zhao and Yoshino [12]. As noted by Zhao and Yoshino [12], the discrepancies may be due to the differentiation of water content; the electrical conductivities of clinopyroxene can be enhanced by the rise of water content. Liu et al. [14] also carried out the impedance spectroscopic measurements on the omphacite single crystals with their correspondent iron content of 7.92 wt% and the water content of 85 ppm wt at pressure of 1.0 GPa and temperature ranges of 623–973 K. The electrical conductivities for omphacite ($10^{-3.6}$–$10^{-1.3}$ S/m) from Liu et al. [14] are higher than our present acquired results, which is possibly caused by the higher contents of iron and water in their measured omphacite. Our obtained electrical conductivities of clinopyroxene are lower than those of natural diopside with iron content of 4.09 wt%, whereas they are higher than synthetic iron-free diopside from Hinze et al. [13]. All of these discrepancies of the electrical conductivities may originate from the distinction of chemical composition in samples, which resemble the effect of iron content on the electrical conductivity of other main rock-forming silicate minerals at high temperature and high pressure [29]. In addition, electrical conductivities of the clinopyroxene–NaCl–H_2O system were measured at temperature ranges of 673–973 K and 1.0 GPa by Sun et al. [15]. Their electrical conductivities are approximately 4–5 orders of magnitude higher than ours, which can be interpreted as the interconnected NaCl-bearing aqueous fluids, and this can greatly enhance the electrical conductivity of fluid-bearing clinopyroxene at high temperature and high pressure.

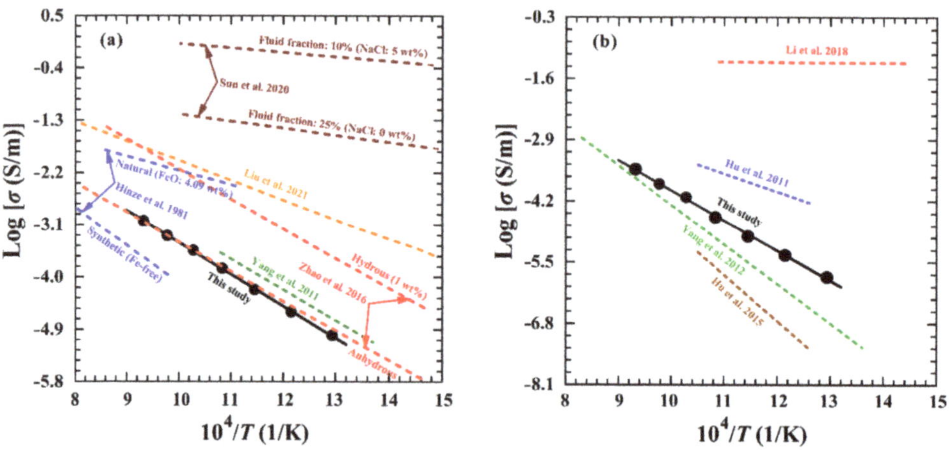

Figure 9. Comparisons of the electrical conductivities of dry clinopyroxene (**a**) and dry plagioclase (**b**) at temperatures of 773–1073 K and 1.0 GPa in present studies with previous results, respectively. The electrical conductivity data for clinopyroxene are from Yang et al. (2011) [11], Zhao et al. (2016) [12] Hinze et al. (1981) [13], Liu et al. (2021) [14] and Sun et al. (2020) [15]. The electrical conductivity data for feldspar are from Yang et al. (2012) [16], Hu et al. (2011, 2015) [17,18], Li et al. (2020) [31].

As another endmember of constituent mineral in gabbro, plagioclase can also alter the chemical composition of hot-pressed sintering samples by changing its corresponding volume percentage, and possibly resulting in the variation of electrical conductivity of gabbro at conditions of high temperature and high pressure. Electrical conductivities of anhydrous plagioclase (its chemical composition is calculated as $An_{57}Ab_{43}$) were investigated under conditions of 773–1073 K and 1.0 GPa, and previously reported results are compared in Figure 9b. It is obvious that our present acquired electrical conductivity of plagioclase (its chemical composition is calculated as $An_{66}Ab_{32}Or_2$) is higher than the previous results reported by Yang et al. [16]. The dominant charge carrier was extrapolated as the monovalent ion of cationic sodium in dry plagioclase and the electrical conductivity might increase with the rise of cationic sodium concentration. The discrepancy between our results and Yang et al. [16] can be ascribed to the higher content of cationic sodium in our samples. The electrical conductivity of albite reported by Hu et al. [17] was higher than our results of plagioclase at 1.0 GPa, which is possibly related to the distinction in the content of cationic sodium. Although the same dominant charge carriers of cationic sodium between our present plagioclase and previously reported albite from Hu et al. [17] were extrapolated, it is evident that our experimental measured plagioclase contains a relatively lower cationic sodium content. In comparison with Hu et al. [18], the electrical conductivity of anhydrous anorthite is lower than our presently obtained results for plagioclase. As usual, the electrical conductivity of dry feldspar is derived from the ionic diffusion of the charge carrier under a designated electric field of alternating current [32]. According to the previous results on the cationic diffusion reported by Behrens et al. [33], the diffusion coefficient of the monovalent cationic sodium is ~3.7 orders of magnitude higher than the bivalent cationic calcium in the ternary structural feldspars at constant temperature. Due to the Nernst–Einstein equation describing the relation between the electrical conductivity and diffusion coefficient, a relatively higher bivalent calcium ion in Hu et al. [18] can result in abnormally lower electrical conductivity results than ours. Besides, Li et al. [31] measured the electrical conductivities of plagioclase–NaCl–water system at temperatures of 400–900 K and 1.2 GPa. The electrical conductivities of NaCl-bearing fluids in plagioclase increase with the rise of the fluid fraction and salinity, which are close to 2–4.5 orders of magnitude higher than those of our anhydrous plagioclase.

Gabbro, as a representative basic intrusive rock, is mainly composed of clinopyroxene and plagioclase. There are significant differences in the mineralogical proportion of gabbro which are outcropped in different regions of oceanic crust and craton crust [5,6,34,35]. Therefore, electrical conductivities of hot-pressed sintering gabbro with various mineralogical proportions ($Cpx_X Pl_{100-X}$, X = 0, 10, 20, 30, 40, 50, 60, 70, 80, 90 and 100 vol%) were systemically investigated under conditions of temperature ranges from 773 K to 1073 K and 1.0 GPa, which were also compared with the available previously reported results on natural gabbro (Figure 10).

As shown in Figure 10, it is clear that our presently acquired electrical conductivities of dry gabbro with various mineralogical proportions are well consistent with the previously reported results. The electrical conductivities of dry natural gabbro in the Oman ophiolite from Saito and Bagdassarov [10] were performed at conditions of temperature ranges of 523–1181 K and pressures of 0.6–0.8 GPa by using the piston-cylinder high-pressure apparatus. As pointed out by Saito et al. [36], the mineralogical composition of natural Oman gabbro is ~63 vol% of plagioclase, ~18 vol% of clinopyroxene, ~12 vol% of orthopyroxene, ~4 vol% of olivine, ~2 vol% of minor amounts of secondary serpentine, and ~1 vol% of magnetite, and the corresponding chemical composition of plagioclase in natural sample is described as $An_{80}Ab_{20}$. In this study, a series of pure hot-pressed sintering anhydrous gabbro samples with the chemical composition of $An_{57}Ab_{43}$ in plagioclase were selected to investigate the effects of mineralogical proportions on the electrical conductivity of samples at high temperature and high pressure. Some feeble discrepancies in the electrical conductivity results of gabbro between Saito and Bagdassarov [10] and ours are possibly related to the complicated mineralogical composition, the occurrence of the inevitable high-temperature dehydration of secondary serpentine, high conductivity phase of magnetite, and chemical composition of plagioclase ($An_{80}Ab_{20}$). Of course, we also find that the electrical conductivity of dry hot-pressed sintering gabbro is relatively lower, which resembles previously reported results on the electrical conductivity of dry silicate minerals [27–29]. When we compared our presently obtained electrical conductivity results with other natural gabbros with high water content reported by Dai et al. [9] and abnormally high content of high-conductivity phases reported by Wang et al. [7] and Bai et al. [8], all of these available discrepancies are possibly due to the differentiation in initial sample composition.

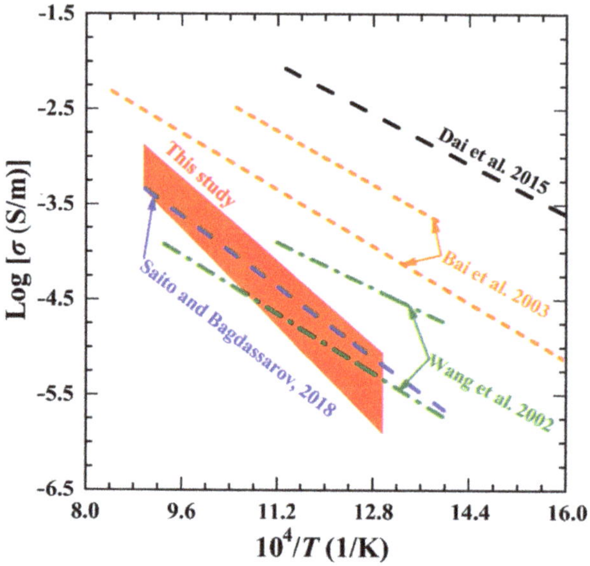

Figure 10. Electrical conductivities of dry hot-pressed sintering gabbro with various mineralogical proportions at the pressure of 1.0 GPa (red shaded area) compared with the previous results.

6. Conduction Mechanism

Since the Arrhenius plot demonstrates a linear relation between the logarithm of electrical conductivity and reciprocal temperature, only one individual conduction mechanism can be identified under various mineralogical compositions ($Cpx_X Pl_{100-X}$, X = 0, 10, 20, 30, 40, 50, 60, 70, 80, 90 and 100 vol%) at the pressure ranges from 1.0 GPa to 3.0 GPa. The present results show that the electrical conductivity of dry hot-pressed gabbro increases with the rise of the volume percentage of clinopyroxene. Previously available reported results demonstrated that the electrical conductivity of dry clinopyroxene depends strongly on iron content and the dominant conduction mechanism is small polaron [11,13,37]. Obviously, the iron content in the hot-pressed sintering gabbro increases with the rise of volume percentage of clinopyroxene, and the electrical conductivities of samples increase with increasing iron content, accordingly. Thus, small polaron is considered to be the dominant conduction mechanism for the anhydrous gabbro. The lattice point defect reactions can be described as follows:

$$Fe_{Mg}^{\times} + h^{\bullet} \rightleftharpoons Fe_{Mg}^{\bullet} \quad (5)$$

where $Fe_{Mg}^{\times}$ and $Fe_{Mg}^{\bullet}$ stand for divalent and trivalent iron ion occupying the site of magnesium ion, and $h^{\bullet}$ stands for a hole, respectively. The main charge carrier of gabbro is the $Fe_{Mg}^{\bullet}$, and it facilitates the conduction by the hopping of the electron between $Fe_{Mg}^{\times}$ and $Fe_{Mg}^{\bullet}$ [9,29,38]. When the nominally anhydrous minerals and rocks contain a trace amount of structural water, it is generally known that some hydrogen-related defects are considered a dominant conduction mechanism [9,27,29]. The FT-IR spectra results indicate that the initial and recovered samples are almost anhydrous. Therefore, the hydrogen-related defects cannot be the dominant conduction mechanism for the anhydrous gabbro. In addition, the ionic conduction is regarded as the main conduction mechanism in some alkali-bearing anhydrous silicate minerals [17,18,39]. It is proposed that the conduction mechanism is the small polaron for dry clinopyroxene and the alkali ion for dry plagioclase [11,15,16,31]. As shown in Figure 8, the electrical conductivity of the dry hot-pressed sintering gabbro shows a negative dependence relation with increasing the volume percentage of plagioclase, and the contribution of plagioclase to electrical conductivity of gabbro can be ruled out. Consequently, the small polaron might be the dominant conduction mechanism in dry hot-pressed sintering gabbro under conditions of high temperature, high pressure, and various mineralogical compositions.

7. Geophysical Implications

As a typical intrusive rock, gabbro is one of the dominant constituent materials in the regions of oceanic crust and craton crust. Petrological and geochemical evidence have demonstrated that the mineralogical proportions of the main constituent minerals in gabbro show an obvious differentiation in various regions [40,41]. Thus, it is crucial to establish the electrical conductivity–depth profile of dry gabbro with various mineralogical proportions on the base of the regions of oceanic crust and craton crust. The relationship between the temperature and depth in the Earth's stationary crust can be obtained by a numerical solution of the heat conduction equation [42],

$$T = T_0 + \left(\frac{Q}{k}\right) Z - \left(\frac{A_0}{2k}\right) Z^2 \tag{6}$$

where T_0 is the surface temperature (K), Q is the surface heat flow (mW/m^2), Z is the lithosphere layer depth (km), k is thermal conductivity (Wm/K) and A_0 is the lithospheric radiogenic heat productivity (μW/m^3). According to the heat conduction equation and thermal calculation parameters, the electrical conductivity–temperature results can be converted to an electrical conductivity–depth profile [25,43,44]. To simplify the conductivity–depth profile model, some necessary assumptions and extrapolations are proposed as follows: (1) the influence of pressure on the electrical conductivity of anhydrous gabbro is ignored; (2) a Ni–NiO solid buffer is adopted to represent the oxygen fugacity in the regions of oceanic crust and craton crust; (3) in the present studies, the hot-pressed sintering gabbro is anhydrous, the effect of water content on the electrical conductivity of sample does not need to be considered; (4) all of the other potentially influential characteristics, including grain size and grain boundary state, on the electrical conductivity of rock can be neglected; (5) the influence of accessory minerals and high conductive phases on the electrical conductivity of gabbro are ignored; (6) dry hot-pressed sintering gabbro with three typical mineralogical proportions ($Cpx_{10}Pl_{90}$, $Cpx_{50}Pl_{50}$, and $Cpx_{90}Pl_{10}$) are adopted to represent the chemical composition of gabbro-rich regions in the oceanic crust and craton crust.

The electrical conductivity–depth profile based on the electrical conductivity of dry hot-pressed sintering gabbro with three typical mineralogical proportions ($Cpx_{10}Pl_{90}$, $Cpx_{50}Pl_{50}$, and $Cpx_{90}Pl_{10}$) for the oceanic crust is constructed at depths of 1–7 km (Figure 11a). Concurrently, we also compared with the previous electrical resistivity results reported by Drury [45]. Two representative geothermal gradient values of 30 K/km and 60 K/km are chosen, which are representing the average and relatively steep plutonic sequence in the oceanic crust, respectively [45–47]. From Figure 11a, it is very obvious that the electrical conductivity of hot-pressed sintering dry gabbro at the geothermal gradient of 60 K/km is far higher than that of 30 K/km at each correspondent typical mineralogical proportions ($Cpx_{10}Pl_{90}$, $Cpx_{50}Pl_{50}$, and $Cpx_{90}Pl_{10}$). At constant geothermal gradients, the electrical conductivity of hot-pressed sintering gabbro will be enhanced ~100 times of magnitude from the $Cpx_{10}Pl_{90}$ to $Cpx_{90}Pl_{10}$ mineralogical proportions. According to the previous results, the electrical conductivity falls within the range of 10^{-4}–10^{-1} S/m in a typical oceanic crust; therefore, the experimental results from the electrical conductivities of the dry hot-pressed sintering gabbro with various mineralogical proportions (Cpx_XPl_{100-X}, X = 0, 10, 20, 30, 40, 50, 60, 70, 80, 90 and 100 vol%) are abnormally lower than the data at depths of 1–7 km.

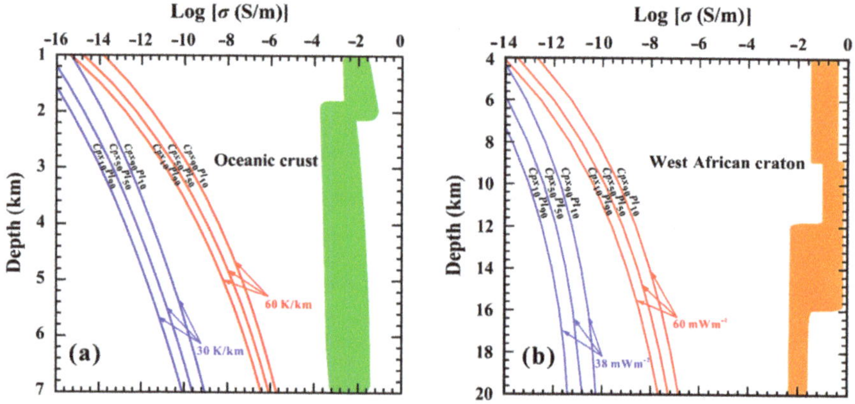

Figure 11. Laboratory-based conductivity–depth profiles constructed from the electrical conductivities of the dry hot-pressed sintering gabbro ($Cpx_{10}Pl_{90}$, $Cpx_{50}Pl_{50,}$ and $Cpx_{90}Pl_{10}$) and the thermodynamic parameters, showing comparison with the conductivity for the anomalous high conductivity zones (HCZ) in the oceanic crust and West African craton, respectively. In (**a**), the red and blue solid lines represent the conductivity–depth profiles on the geothermal gradients of 60 K/km and 30 K/km, respectively. In (**b**), the red solid lines and the blue solid lines are based on the surface heat flow values of 60 mW/m^2 and 38 mW/m^2, respectively. The green region represents the oceanic crust with a high conductivity anomaly in (**a**), and the MT data derived from a high-conductivity anomaly in the West African craton are shown as the orange region in (**b**) (reproduced with permission from the reference of Le Pape et al. [4], published by Elsevier, 2017; reproduced with permission from the reference of Drury [45], published by Oxford University Press, 1979).

At the same time, two typical heat flow values of 38 and 60 mW/m^2 were selected for the West African craton [48,49]. As shown in Figure 11b, the electrical conductivity–depth profile was successfully established from the West African craton at depths of 4–20 km, and the magnetotelluric conductivity results from the West African craton are also included [4]. It is evident that the electrical conductivity of dry hot-pressed sintering gabbro at the heat flow value of 60 mW/m^2 is far higher than that of 38 mW/m^2 at each correspondent typical mineralogical proportions ($Cpx_{10}Pl_{90}$, $Cpx_{50}Pl_{50,}$ and $Cpx_{90}Pl_{10}$). At constant heat flow values, the electrical conductivity of hot-pressed sintering gabbro will be enhanced ~1000 times of magnitude from the $Cpx_{10}Pl_{90}$ to $Cpx_{90}Pl_{10}$ mineralogical proportions. Based on the previous results from the field magnetotelluric data, the electrical conductivities are within the range of 10^{-2}–10^{0} S/m in West African craton. Consequently, the experimental results from the electrical conductivities of the dry hot-pressed sintering gabbro with various mineralogical proportions (Cpx_XPl_{100-X}, X = 0, 10, 20, 30, 40, 50, 60, 70, 80, 90 and 100 vol%) are abnormally lower than field magnetotelluric data at depths of 4–20 km.

In conclusion, our obtained experimental results on the electrical conductivities of the dry hot-pressed sintering gabbro with various mineralogical proportions (Cpx_XPl_{100-X}, X = 0, 10, 20, 30, 40, 50, 60, 70, 80, 90 and 100 vol%) at high temperature and high pressure are abnormally lower than those of field magnetotelluric data of the gabbro-rich regions, which cannot explain the high conductivity anomalies for the oceanic crust and West African craton. Some potential causes need to be considered conscientiously, such as the structural water in nominally anhydrous minerals [9,28,30,50,51], dehydration (oxidation–dehydrogenation) of hydrous minerals [23,52,53], partial melting [54–57], the interconnected high conductive phases [24,28,58,59], and the salinity-bearing (or water-bearing) fluids [15,31,60–62], in the future. Although the present acquired electrical conductivity results on the dry hot-pressed sintering gabbro with various mineralogical proportions cannot explain the high conductivity anomaly in the oceanic crust and West African craton, it can provide one reasonable constraint on the mineralogical composition in these representative gabbro-rich regions.

Author Contributions: L.D. and H.H.: designing the project; M.W., L.D. and H.H.: writing the initial draft of the work and the final paper, as well as interpreting the results; M.W. and W.S.: performing electrical conductivity experiments; M.W., Z.H. and C.J.: performing and interpreting scanning electron microscope analyses; All authors discussed the results and commented on the manuscript. All authors have read and agreed to the published version of the manuscript.

Funding: This research was financially supported by the NSF of China (42072055, 41774099, 41772042 and 42002043), Youth Innovation Promotion Association of CAS (Grant No.2019390) and Special Fund of the West Light Foundation of CAS (2018).

Data Availability Statement: The data that support the findings of this study are available from the corresponding author upon reasonable request.

Acknowledgments: We thank the editor of Julien Siebert for kindly handling our paper, as well as three anonymous reviewers for their constructive and enlightened advice in the revising process.

Conflicts of Interest: The authors declare no conflict of interest.

References

1. Baba, K. Electrical structure in marine tectonic settings. *Surv. Geophys.* **2005**, *26*, 701–731. [CrossRef]
2. Kaya, T.; Kasaya, T.; Tank, S.B.; Ogawa, Y.; Tuncer, M.K.; Oshiman, N.; Honkura, Y.; Matsushima, M. Electrical characterization of the north anatolian fault zone underneath the Marmara Sea, turkey by ocean bottom magnetotellurics. *Geophys. J. Int.* **2013**, *193*, 664–677. [CrossRef]
3. Usui, Y.; Kasaya, T.; Ogawa, Y.; Iwamoto, H. Marine magnetotelluric inversion with an unstructured tetrahedral mesh. *Geophys. J. Int.* **2018**, *214*, 952–974. [CrossRef]
4. Le Pape, F.; Jones, A.G.; Jessell, M.W.; Perrouty, S.; Gallardo, L.A.; Baratoux, L.; Hogg, C.; Siebenaller, L.; Toure, A.; Ouiya, P.; et al. Crustal structure of southern Burkina Faso inferred from magnetotelluric, gravity and magnetic data. *Precambrian Res.* **2017**, *300*, 261–272. [CrossRef]
5. Ringwood, A.E.; Green, D.H. An experimental investigation of the gabbro-eclogite transformation and some geophysical implications. *Tectonophysics* **1966**, *3*, 383–427. [CrossRef]
6. Christensen, N.I.; Salisbury, M.H. Structure and constitution of the lower oceanic crust. *Rev. Geophys.* **1975**, *13*, 57–86. [CrossRef]
7. Wang, D.J.; Liu, C.Q.; Li, H.P.; Yi, L.; Su, G.L.; Ding, D.Y. Impedance spectra of hot, dry gabbro at high temperature and pressure. *Prog. Nat. Sci.* **2002**, *12*, 397–400.
8. Bai, L.P.; Du, J.G.; Liu, W.; Zhou, W.G. Experimental studies of electrical conductivities and P-wave velocities of gabbro at high pressures and high temperatures. *Sci. China Ser. D* **2003**, *46*, 895–908. [CrossRef]
9. Dai, L.D.; Hu, H.Y.; Li, H.P.; Hui, K.S.; Jiang, J.J.; Li, J.; Sun, W.Q. Electrical conductivity of gabbro: The effects of temperature, pressure and oxygen fugacity. *Eur. J. Mineral.* **2015**, *27*, 215–224. [CrossRef]
10. Saito, S.; Bagdassarov, N.S. Laboratory measurements of electrical conductivity in a gabbro of the Oman ophiolite at high–pressures and high–temperatures: Implications for interpretation of resistivity structures of lower oceanic crust. *J. Miner. Petrol. Sci.* **2018**, *113*, 112–117. [CrossRef]
11. Yang, X.Z.; Keppler, H.; McCammon, C.; Ni, H.W.; Xia, Q.K.; Fan, Q.C. Effect of water on the electrical conductivity of lower crustal clinopyroxene. *J. Geophys. Res. Solid Earth* **2011**, *116*, B04208. [CrossRef]
12. Zhao, C.C.; Yoshino, T. Electrical conductivity of mantle clinopyroxene as a function of water content and its implication on electrical structure of uppermost mantle. *Earth Planet. Sci. Lett.* **2016**, *447*, 1–9. [CrossRef]
13. Hinze, E.; Will, G.; Cemic, L. Electrical conductivity measurements on synthetic olivines and on olivine, enstatite and diopside from Dreiser Weiher, Eifel (Germany) under defined thermodynamic activities as a function of temperature and pressure. *Phys. Earth Planet. Inter.* **1981**, *25*, 245–254. [CrossRef]
14. Liu, H.Y.; Zhang, K.; Ingrin, J.; Yang, X.Z. Electrical conductivity of omphacite and garnet indicates limited deep water recycling by crust subduction. *Earth Planet. Sci. Lett.* **2021**, *559*, 116784. [CrossRef]
15. Sun, W.Q.; Dai, L.D.; Li, H.P.; Hu, H.Y.; Jiang, J.J.; Wang, M.Q. Electrical conductivity of clinopyroxene–NaCl–H_2O system at high temperatures and pressures: Implications for high-conductivity anomalies in the deep crust and subduction zone. *J. Geophys. Res. Solid Earth* **2020**, *125*, e2019JB019093. [CrossRef]
16. Yang, X.Z.; Keppler, H.; McCammon, C.; Ni, H.W. Electrical conductivity of orthopyroxene and plagioclase in the lower crust. *Contrib. Mineral. Petrol.* **2012**, *163*, 33–48. [CrossRef]
17. Hu, H.Y.; Li, H.P.; Dai, L.D.; Shan, S.M.; Zhu, C.M. Electrical conductivity of albite at high temperatures and high pressures. *Am. Mineral.* **2011**, *96*, 1821–1827. [CrossRef]
18. Hu, H.Y.; Dai, L.P.; Li, H.P.; Hui, K.S.; Li, J. Temperature and pressure dependence of electrical conductivity in synthetic anorthite. *Solid State Ionics* **2015**, *276*, 136–141. [CrossRef]

19. Yang, X.Z.; Heidelbach, F. Grain size effect on the electrical conductivity of clinopyroxene. *Contrib. Mineral. Petrol.* **2011**, *163*, 939–947. [CrossRef]
20. Bell, D.R.; Ihinger, P.D.; Rossman, G.R. Quantitative analysis of trace OH in garnet and pyroxenes. *Am. Mineral.* **1995**, *80*, 465–474. [CrossRef]
21. Dai, L.D.; Hu, H.Y.; Li, H.P.; Wu, L.; Hui, K.S.; Jiang, J.J.; Sun, W.Q. Influence of temperature, pressure, and oxygen fugacity on the electrical conductivity of dry eclogite, and geophysical implications. *Geochem. Geophys. Geosyst.* **2016**, *17*, 2394–2407. [CrossRef]
22. Dai, L.D.; Li, H.P.; Hu, H.Y.; Shan, S.M. Experimental study of grain boundary electrical conductivities of dry synthetic peridotite under high-temperature, high-pressure, and different oxygen fugacity conditions. *J. Geophys. Res. Solid Earth* **2008**, *113*, B12211. [CrossRef]
23. Hu, H.Y.; Dai, L.D.; Li, H.P.; Sun, W.Q.; Li, B. Effect of dehydrogenation on the electrical conductivity of Fe-bearing amphibole and its implications for the high conductivity anomalies in subduction zones and continental crust. *Earth Planet. Sci. Lett.* **2018**, *498*, 27–37. [CrossRef]
24. Dai, L.D.; Hu, H.Y.; Sun, W.Q.; Li, H.P.; Liu, C.C.; Wang, M.Q. Influence of high conductive magnetite impurity on the electrical conductivity of dry olivine aggregates at high temperature and high pressure. *Minerals* **2019**, *9*, 44. [CrossRef]
25. Dai, L.D.; Li, H.P.; Hu, H.Y.; Shan, S.M.; Jiang, J.J.; Hui, K.S. The effect of chemical composition and oxygen fugacity on the electrical conductivity of dry and hydrous garnet at high temperatures and pressures. *Contrib. Mineral. Petrol.* **2012**, *163*, 689–700. [CrossRef]
26. Dai, L.D.; Li, H.P.; Hu, H.Y.; Jiang, J.J.; Hui, K.S.; Shan, S.M. Electrical conductivity of $Alm_{82}Py_{15}Grs_3$ almandine-rich garnet determined by impedance spectroscopy at high temperatures and high pressures. *Tectonophysics* **2013**, *608*, 1086–1093. [CrossRef]
27. Dai, L.D.; Karato, S.I. Electrical conductivity of orthopyroxene: Implications for the water content of the asthenosphere. *Proc. Jpn. Acad. B* **2009**, *85*, 466–475. [CrossRef]
28. Dai, L.D.; Karato, S.I. Electrical conductivity of wadsleyite at high temperatures and high pressures. *Earth Planet. Sci. Lett.* **2009**, *287*, 277–283. [CrossRef]
29. Dai, L.D.; Karato, S.I. Influence of FeO and H on the electrical conductivity of olivine. *Phys. Earth Planet. Inter.* **2014**, *237*, 73–79. [CrossRef]
30. Dai, L.D.; Karato, S.I. High and highly anisotropic electrical conductivity of the asthenosphere due to hydrogen diffusion in olivine. *Earth Planet. Sci. Lett.* **2014**, *408*, 79–86. [CrossRef]
31. Li, P.; Guo, X.Z.; Chen, S.B.; Wang, C.; Yang, J.L.; Zhou, X.F. Electrical conductivity of the plagioclase–NaCl–water system and its implication for the high conductivity anomalies in the mid-lower crust of Tibet Plateau. *Contrib. Mineral. Petrol.* **2018**, *173*, 16. [CrossRef]
32. Dai, L.D.; Hu, H.Y.; Li, H.P.; Sun, W.Q.; Jiang, J.J. Influence of anisotropy on the electrical conductivity and diffusion coefficient of dry K-feldspar: Implications for the mechanism of conduction. *Chin. Phys. B* **2018**, *27*, 028703. [CrossRef]
33. Behrens, H.; Johannes, W.; Schmalzried, H. On the mechanisms of cation diffusion processes in ternary feldspars. *Phys. Chem. Minerals* **1990**, *17*, 62–78. [CrossRef]
34. Boudier, F.; Nicolas, A.; Ildefonse, B. Magma chambers in the Oman ophiolite: Fed from the top and the bottom. *Earth Planet. Sci. Lett.* **1996**, *144*, 239–250. [CrossRef]
35. Ilboudo, H.; Sawadogo, S.; Traoré, A.S.; Sama, M.; Wenmenga, U.; Lompo, M. Intrusion-related gold mineralization: Inata gold deposit, Bélahourou district, Northern Burkina Faso (West-Africa). *J. Afr. Earth Sci.* **2018**, *148*, 52–58. [CrossRef]
36. Saito, S.; Ishikawa, M.; Arima, M.; Tatsumi, Y. Laboratory measurements of 'porosity–free' intrinsic *Vp* and *Vs* in an olivine gabbro of the Oman ophiolite: Implication for interpretation of the seismic structure of lower oceanic crust. *Island Arc.* **2015**, *24*, 131–144. [CrossRef]
37. Huebner, J.S.; Voigt, D.E. Electrical conductivity of diopside: Evidence for oxygen vacancies. *Am. Mineral.* **1988**, *73*, 1235–1254.
38. Poe, B.T.; Romano, C.; Varchi, V.; Misiti, V.; Scarlato, P. Electrical conductivity of a phonotephrite from Mt. Vesuvius: The importance of chemical composition on the electrical conductivity of silicate melts. *Chem. Geol.* **2008**, *256*, 193–202. [CrossRef]
39. Hu, H.Y.; Li, H.P.; Dai, L.D.; Shan, S.M.; Zhu, C.M. Electrical conductivity of alkali feldspar solid solutions at high temperatures and high pressures. *Phys. Chem. Minerals* **2013**, *40*, 51–62. [CrossRef]
40. Wiebe, R.A. The pleasant bay layered gabbro-diorite, coastal maine: Ponding and crystallization of basaltic injections into a silicic magma chamber. *J. Petrol.* **1993**, *34*, 461–489. [CrossRef]
41. Khromykh, S.V.; Izokh, A.E.; Gurova, A.V.; Cherdantseva, M.V.; Savinsky, I.A.; Vishnevsky, A.V. Syncollisional gabbro in the Irtysh shear zone, Eastern Kazakhstan: Compositions, geochronology, and geodynamic implications. *Lithos* **2019**, *346–347*, 105144. [CrossRef]
42. Selway, K.; Yi, J.; Karato, S.I. Water content of the Tanzanian lithosphere from magnetotelluric data: Implications for cratonic growth and stability. *Earth Planet. Sci. Lett.* **2014**, *388*, 175–186. [CrossRef]
43. Dai, L.D.; Hu, H.Y.; Li, H.P.; Jiang, J.J.; Hui, K.S. Influence of temperature, pressure, and chemical composition on the electrical conductivity of granite. *Am. Mineral.* **2014**, *99*, 1420–1428. [CrossRef]
44. Sun, W.Q.; Dai, L.D.; Li, H.P.; Hu, H.Y.; Liu, C.C.; Wang, M.Q. Effect of temperature, pressure and chemical compositions on the electrical conductivity of schist: Implications for electrical structures under the Tibetan plateau. *Materials* **2019**, *12*, 961. [CrossRef] [PubMed]

45. Drury, M.J. Electrical resistivity models of the oceanic crust based on laboratory measurements on basalts and gabbros. *Geophys. J. R. Astron. Soc.* **1979**, *56*, 241–253. [CrossRef]
46. Hyndman, R.D.; Drury, M.J. The physical properties of oceanic basement rocks from deep drilling on the mid-Atlantic ridge. *J. Geophys. Res.* **1976**, *81*, 4042–4052. [CrossRef]
47. Nehlig, P.; Juteau, T. Flow porosities, permeabilities and preliminary data on fluid inclusions and fossil thermal gradients in the crustal sequence of the Sumail ophiolite (Oman). *Tectonophysics* **1988**, *151*, 199–221. [CrossRef]
48. Lucazeau, F.; Lesquer, A.; Vasseur, G. *Trends of Heat Flow Density from West Africa. Terrestrial Heat Flow and the Lithosphere Structure. Exploration of the Deep Continental Crust*; Čermák, V., Rybach, L., Eds.; Springer: Berlin, Germany, 1991; pp. 417–425.
49. Le Pape, F.; Jones, A.G.; Jessell, M.W.; Hogg, C.; Siebenaller, L.; Perrouty, S.; Touré, A.; Ouiya, P.; Boren, G. The nature of the southern West African craton lithosphere inferred from its electrical resistivity. *Precambrian Res.* **2021**, *358*, 106190. [CrossRef]
50. Dai, L.D.; Hu, H.Y.; He, Y.; Sun, W.Q. Some new progress in the experimental measurements on electrical property of main minerals in the upper mantle at high temperatures and high pressures. In *Mineralogy*; René, M., Ed.; IntechOpen: London, UK, 2022; *in press*. [CrossRef]
51. Dai, L.D.; Karato, S.I. Electrical conductivity of Ti-bearing hydrous olivine aggregates at high temperature and high pressure. *J. Geophys. Res. Solid Earth* **2020**, *125*, e2020JB020309. [CrossRef]
52. Manthilake, G.; Mookherjee, M.; Bolfan-Casanova, N.; Andrault, D. Electrical conductivity of lawsonite and dehydrating fluids at high pressures and temperatures. *Geophys. Res. Lett.* **2015**, *42*, 7398–7405. [CrossRef]
53. Hu, H.Y.; Dai, L.D.; Li, H.P.; Hui, K.S.; Sun, W.Q. Influence of dehydration on the electrical conductivity of epidote and implications for high conductivity anomalies in subduction zones. *J. Geophys. Res. Solid Earth* **2017**, *122*, 2751–2762. [CrossRef]
54. Maumus, J.; Bagdassarov, N.; Schmeling, H. Electrical conductivity and partial melting of mafic rocks under pressure. *Geochim. Cosmochim. Acta* **2005**, *69*, 4703–4718. [CrossRef]
55. Pommier, A.; Leinenweber, K.; Kohlstedt, D.L.; Qi, C.; Garnero, E.J.; Mackwell, S.J.; Tyburczy, J.A. Experimental constraints on the electrical anisotropy of lithosphere-asthenosphere system. *Nature* **2015**, *522*, 202–206. [CrossRef] [PubMed]
56. Guo, X.; Zhang, L.; Su, X.; Mao, Z.; Gao, X.Y.; Yang, X.Z.; Ni, H.W. Melting inside the Tibetan crust? Constraint from electrical conductivity of peraluminous granitic melt. *Geophys. Res. Lett.* **2018**, *45*, 3906–3913. [CrossRef]
57. Freitas, D.; Manthilake, G.; Chantel, J.; Bouhifd, M.A.; Andrault, D. Simultaneous measurements of electrical conductivity and seismic wave velocity of partially molten geological materials: Effect of evolving melt texture. *Phys. Chem. Minerals* **2019**, *46*, 535–551. [CrossRef]
58. Jödicke, H.; Nover, G.; Kruhl, J.H.; Markfort, R. Electrical properties of a graphite-rich quartzite from a former lower continental crust exposed in the Serre San Bruno, Calabria (southern Italy). *Phys. Earth Planet. Inter.* **2007**, *165*, 56–67. [CrossRef]
59. Sun, W.Q.; Jiang, J.J.; Dai, L.D.; Hu, H.Y.; Wang, M.Q.; Qi, Y.Q.; Li, H.P. Electrical properties of dry polycrystalline olivine mixed with various chromite contents: Implications for the high conductivity anomalies in subduction zones. *Geosci. Front.* **2021**, *12*, 101178. [CrossRef]
60. Huang, Y.; Guo, H.; Nakatani, T.; Uesugi, K.; Nakamura, M.; Keppler, H. Electrical conductivity in texturally equilibrated fluid-bearing forsterite aggregates at 800 °C and 1 GPa: Implications for the high electrical conductivity anomalies in mantle wedges. *J. Geophys. Res. Solid Earth* **2021**, *126*, e2020JB021343. [CrossRef]
61. Hu, H.Y.; Dai, L.D.; Sun, W.Q.; Zhuang, Y.K.; Liu, K.X.; Yang, L.F.; Pu, C.; Hong, M.L.; Wang, M.Q.; Hu, Z.M.; et al. Some remarks on the electrical conductivity of hydrous silicate minerals in the Earth crust, upper mantle and subduction zone at high temperatures and high pressures. *Minerals* **2022**, *12*, 161. [CrossRef]
62. Sun, W.Q.; Jiang, J.J.; Dai, L.D.; Hu, H.Y.; Wang, M.Q.; Qi, Y.Q.; Li, H.P. Influence of saline fluids on the electrical conductivity of olivine aggregates at High temperature and high pressure and its geological implications. *Front. Earth Sci.* **2021**, *9*, 749896. [CrossRef]

minerals

MDPI

Brief Report

Study on the Phase Transition from Quartz to Coesite under High Temperature and High Pressure

Dongsheng Ren

State Key Laboratory of Earthquake Dynamics, Institute of Geology, China Earthquake Administration, Beijing 100029, China; rendongsheng@ies.ac.cn

Abstract: Quartz is an important component of the Earth. In this study, experiments were conducted at temperatures between 600 to 700 °C, confining pressures between 1.5 and 1.8 GPa, and differential stress conditions. It was found that coesite production is closely related to differential stress, reaction time, and reaction temperature, with coesite formation being a multifactorial coupling process.

Keywords: high temperature; high pressure; differential stress; coesite; phase transition

Citation: Ren, D. Study on the Phase Transition from Quartz to Coesite under High Temperature and High Pressure. *Minerals* **2022**, *12*, 963. https://doi.org/10.3390/min12080963

Academic Editors: Lidong Dai, Haiying Hu, Jianjun Jiang and Jordi Ibanez-Insa

Received: 15 March 2022
Accepted: 19 July 2022
Published: 29 July 2022

1. Introduction

Quartz is an important component of the Earth's interior, and the study of its phase transitions under high pressure is crucial to understanding the stability of various mineral assemblages, and the temperature and pressure environment deep in the Earth's interior, and to determining the diagenesis depth of certain rocks containing high-pressure mineral inclusions. In laboratory studies, the quartz phase transitions to coesite, then stishovite at high temperatures and pressures, and the super-stishovite phase has a $CaCl_2$-type and α-PbO_2-type structure under ultra-high pressure [1–3].

Since the temperature and pressure conditions required for the phase transition of quartz into coesite are equivalent to those in the Earth's interior at a depth of about 100 km, the coesite found near the Earth's surface may have been brought to the surface by geological activity after forming at such depths. In addition to the above foldback hypothesis [4], the formation of high-pressure quartz phases may also be caused by shock and differential stress effects [5–9]. It has been demonstrated that coesite can occur at a confining pressure of 1 to 2 GPa under conditions of differential stress and large strain [10,11]. Hirth and Tullis (1994) investigated the brittle–plastic transformation of quartz using a piston-cylinder pressure apparatus and found that stable coesite is observed at low confining pressures (1.2 to 1.25 GPa) in the presence of relatively large differential stresses and medium strains [12]. The same conclusion was reached in high-temperature and high-pressure experiments [1]. Richter (2016) [13] identified coesite formation by shear experiments at 600 to 900 °C and 1 to 1.5 GPa. These studies indicate that coesite formation under differential stress conditions requires a lower confining pressure than what is necessary to stabilize coesite under hydrostatic pressure (>3 GPa) [13]. However, the above studies focused on the important role of the maximum principal stress. They believe that coesite can appear if the maximum principal stress exceeds the phase transition pressure while being under hydrostatic pressure at the same temperature.

In this study, we conducted experiments at 600 to 700 °C and 1.5 to 1.8 GPa using the Griggs apparatus to investigate the relationship between the quartz–coesite phase transition and temperature, phase transition time, strain, and differential stress, which is of great significance.

2. Experimental Materials and Methods

2.1. Starting Materials

The α-quartz powder used in these experiments was purchased from Aladdin Co., Ltd. in Shanghai, China. It had a density of 2.65 g/cm^3, a purity of 99%, and a particle size of 250 μm. The α-quartz was dried thoroughly in an oven at 150 °C. A sample of 0.850 g of powder was weighed on an analytical balance and wrapped with silver foil, then placed in a ϕ10 press mold and pressed to 2 MPa using a powder pressing machine to obtain a cylinder with a length of 20 ± 0.05 mm and a diameter of 10 ± 0.05 mm.

2.2. Experimental Design and Process

The sample assembly of the experiment is shown in Figure 1; this machine is in the Institute of Geology, China Earthquake Administration [1]. The axial pressure of this equipment can be controlled by load and displacement. First, we controlled the load by applying a small pressure (about 0.5 GPa) to the sample, as to make the components in the sample have close contact. At a lower pressure, we heated the sample to the target experimental temperature at a heating rate of 10 °C/min. Then, we continued to use the load control method to load the pressure on the sample to the target confining pressure. Finally, we controlled the axial pressure through displacement loading with the loading speed of 5 × 10^{-6} m/s; after loading to the target pressure value, the pressure was maintained. The schematic diagram of pressure loading and heating process is shown in Figure 2. The value of the confining pressure was obtained through pressure calibration, the axial pressure was calculated according to the axial pressure load, and the strain was calculated through the axial displacement sensor. After the experiment, the heating power supply was cut off, and the axial pressure and confining pressure were quickly unloaded to make the sample quench quickly. Then, the samples were removed and analyzed by Raman spectroscopy to determine if any phase transitions had occurred. Raman spectroscopy was completed at the Institute of Geochemistry, Chinese Academy of Sciences. For details, see [14]. The samples used for the Raman spectrum test were selected from the products after the experiment, and the specific position was selected on the upper surface, close to the corundum column (No. 7 in Figure 1). The reason for choosing this place is that the differential stress environment here is conducive to the production of coesite. The pressure environments of other parts of the sample consist of quasi-hydrostatic pressure. If coesite does not appear at the selected test location, it is almost impossible to appear elsewhere.

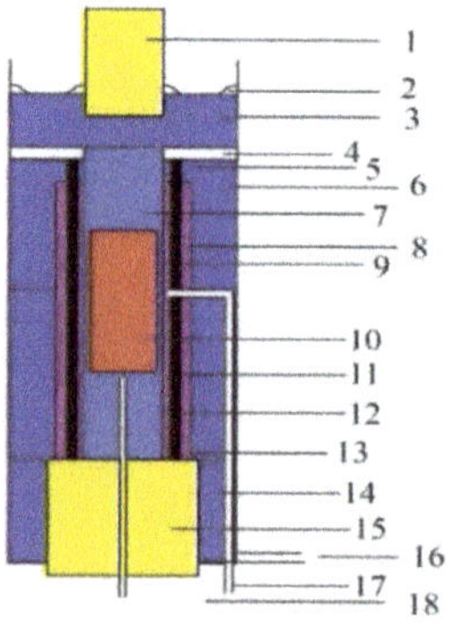

Figure 1. The α-quartz phase transition experimental assembly. (1) Tungsten carbide shaft rod, (2) confining pressure piston sealing ring, (3) lead ring, (4) copper gasket, (5) upper conductive copper ring, (6) pressure-transmitting medium salt sleeve, (7) corundum column, (8) outer pyrophyllite sleeve, (9) sample sealing sleeve, (10) sample, (11) graphite furnace, (12) inner pyrophyllite sleeve, (13) lower conductive copper ring, (14) lower pyrophyllite sleeve, (15) tungsten carbide briquette, (16) insulating mica sheet, (17) thermocouple, (18) sample bottom thermocouple.

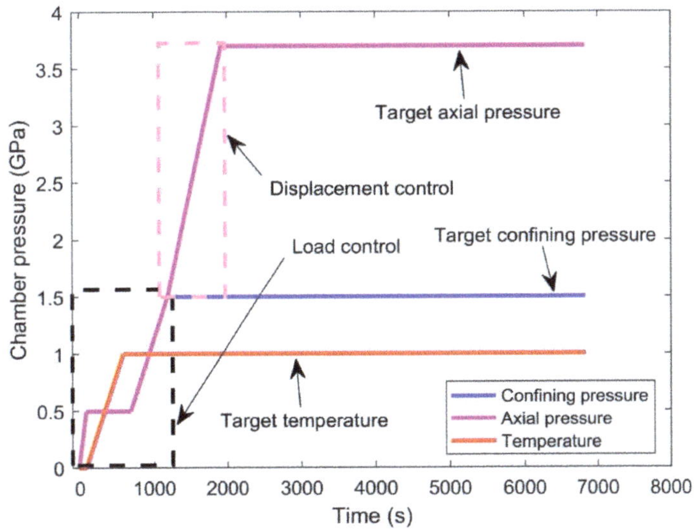

Figure 2. Pressure loading and heating process during the experiment.

3. Results and Discussion

A total of five sets of experiments were conducted in this study, as shown in Table 1. The resulting samples were sectioned for Raman spectroscopy analysis. Experiments 1 and 3 did not reveal coesite being formed, whereas Experiments 2, 4, and 5 found coesite at the corundum indenter. Figure 3 shows the Raman spectra of the starting materials and the resulting products from Experiments 2, 4, and 5.

Table 1. Experimental conditions of the quartz phase transition under differential stress.

	Temperature/°C	Confining Pressure/GPa	Maximum Principal Stress/GPa	Differential Stress/GPa	Mean Stress/GPa	Strain	Holding Time/h	Coesite Present (Y/N)
1	700	1.5	3.7	2.2	2.0	72.1	36	N
2	700	1.6	3.9	2.3	2.1	72.3	36	Y
3	600	1.8	4.1	2.3	2.4	73.8	24	N
4	600	1.8	4.2	2.4	2.4	75.6	36	Y
5	700	1.7	4.0	2.3	2.3	71.1	24	Y

The experimental data presented in Table 1 and previous experimental data obtained under similar experimental conditions are plotted on the quartz–coesite phase transition diagram under hydrostatic pressure (Figure 4) [1,11]. As observed, the results from this experimental study are close to those of Hirth (1994) [12] and differ somewhat from those of Zhou et al. (2005) [1]. In Zhou et al. (2005) [1], except for the cultellation of the maximum principal stress, which is close to the quartz–coesite phase transition line under hydrostatic pressure, the cultellation of the confining pressure, differential stress, and average stress were much lower than the quartz–coesite phase transition line under hydrostatic pressure; the experimental temperature was about 1000 °C, which is very high. According to the current study and Hirth's findings, only the cultellation of the confining pressure is much lower than the quartz–coesite phase transition line under hydrostatic pressure, as the cultellation of the maximum principal stress is much higher than the quartz–coesite phase transition line under hydrostatic pressure, and the cultellation of the differential stress and average stress is distributed near the quartz–coesite phase transition line. However, the reaction temperatures in this study were 600 and 700 °C, which is relatively low.

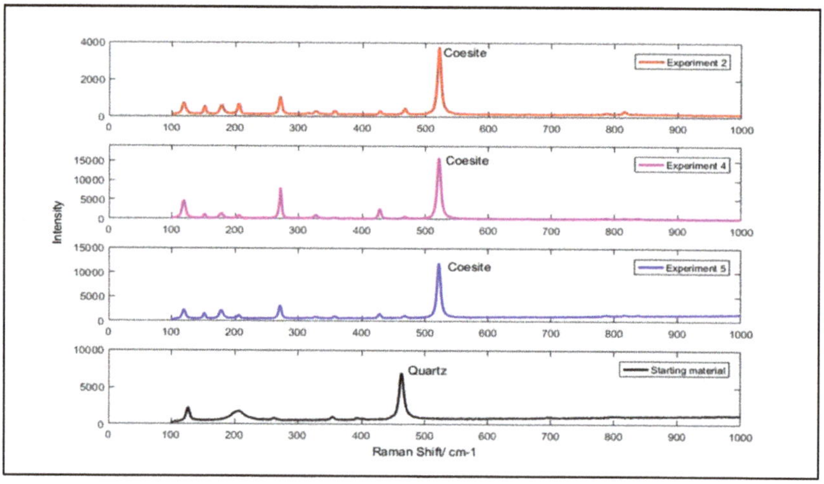

Figure 3. Raman spectra of the raw material and the products after the experiments that resulted in coesite forming.

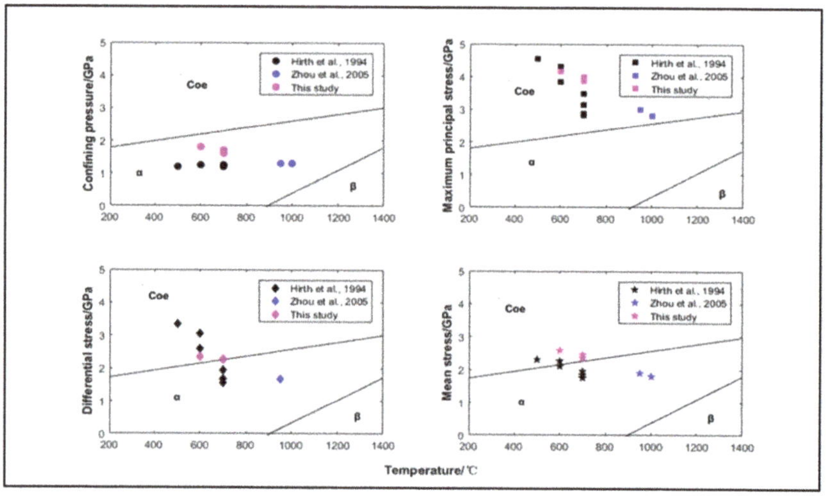

Figure 4. Pressure and stress conditions as a function of temperature for different phases of quartz. Coe—coesite [1,12].

For Experiments 1 and 2, the reaction temperature and pressure holding time are the same, but the maximum principal stress and differential stress in Experiment 2 are greater than those in Experiment 1. No coesite developed in Experiment 1, but it did develop in Experiment 2. Therefore, it is believed that the maximum principal stress and differential stress are the key factors behind coesite formation in this case. This view is also consistent with that of Hirth (1994) [12]. Meanwhile, we found that the coesite production is also related to the pressure holding time, and a shorter pressure holding time is not conducive to the production of coesite, which is derived from comparing the results of Experiments 3 and 4. In these experiments, the reaction temperature and stress conditions are similar but the pressure holding time differs, with no coesite developing in Experiment 3, although it did develop in Experiment 4. In addition, we also found that the higher temperature in this deformation region favors the emergence of coesite, which we concluded from comparing the results of Experiments 3 and 5. The comparison between the experimental results of Zhou et al. (2005) [1] and our results and those of Hirth (1994) [12] also shows the temperature effect on coesite formation.

In Zhou et al. (2005) [1], the large deformation is also a major factor in coesite generation. In this study, the quartz deformation is also very large, with the strain exceeding 70%; thus, we believe that a large deformation is one of the main reasons for the development of coesite.

All of the studies mentioned above (Zhou et al., 2005 [1]; Hirth, 1994 [12]; and this study) indicate that the phase transformation of quartz to coesite can occur at a certain temperature with a small sample confining pressure and high differential stress. However, it is believed that the emergence of coesite is a complex process, which is not dominated by a single factor but involves multifactorial coupling, and that the quartz–coesite phase transition is only observed at a specific temperature (500 °C) and under high differential stress with a long reaction time.

4. Conclusions

In this study, experiments were conducted over the temperature range of 600 to 700 °C, confining pressures of 1.5 to 1.8 GPa, and differential stress conditions. It was found that the generation of coesite is closely related to differential stress, reaction time, and reaction temperature, with coesite formation being the result of a multifactorial coupling process. Compared with previous studies (Hirth, 1994 [12]; Zhou et al., 2005 [1]), the highlight of our experiment is the discovery of coesite under low confining pressure and large strain conditions. However, there are some limitations in these results. Current studies have been restricted to the qualitative discussion of coesite formation conditions, but no quantitative experimental results are available to help explain the relationships between coesite formation and differential stress.

Funding: This research was funded by the National Natural Science Foundation of China, grant number 42104176.

Data Availability Statement: Not applicable.

Conflicts of Interest: The author declares no conflict of interest.

References

1. Zhou, Y.; He, C.; Ma, S.; Song, J.; Ma, J. Effect of differential stress on quartz coesite conversion pressure. *Sci. Bull.* **2005**, *50*, 565–569. [CrossRef]
2. Einicke, W.; Enke, D.; Dvoyashkin, M.; Valiullin, R.; Gläser, R. The Mechanism of pseudomorphic transformation of spherical silica gel into mcm-41 studied by PFG NMR diffusometry. *Materials* **2013**, *6*, 3688–3709. [CrossRef] [PubMed]
3. Klemm, D.H.W.; Prieto, D.M.J.; Xiong, D.F.; Hassine, G.B.; Heyde, M.; Menzel, D.; Freund, H.J. A Silica Bilayer Supported on Ru (0001): Following the Crystalline-to Vitreous Transformation in Real Time with Spectro-microscopy. *Angew. Chem.* **2020**, *63*, 813–820.
4. Wain, A. New evidence for coesite in eclogite and gneisses: Defining an ultrahigh-pressure province in the Western Gneiss region of Norway. *Geology* **1997**, *25*, 927–930. [CrossRef]
5. Spektor, K.; Nylen, J.; Mathew, R.; Edén, M.; Stoyanov, E.; Navrotsky, A.; Häussermann, U. Formation of Hydrous Stishovite from Coesite in High-Pressure Hydrothermal Environments. *Mineral. Soc. Am.* **2016**, *101*, 2514–2524. [CrossRef]
6. Zhou, W.G.; Xie, H.S.; Zhao, Z.D.; Zhou, H.; Guo, J. Calculation of the Strain, Stress and Elastic Energy for α–β Quartz Transition and Its Geological Significance. *J. Geophys.* **2002**, *16*, 241–248.
7. Chen, X.Y.; Feng, W.J.; Zhang, G.Y.; Gao, Y. Raman Spectra of Quartz and Pb^{4+} SiO_2 Crystals at Different Temperature and Pressure. *Crystals* **2019**, *9*, 569. [CrossRef]
8. Griggs, D. Hydrolytic Weakening of Quartz and Other Silicates. *Phys. Chem. Miner.* **1967**, *14*, 19–31. [CrossRef]
9. Hülsmans, A.; Schmücker, M.; Mader, W.; Schneider, H. The Transformation of Andalusite to Mullite and Silica: Part I. Transformation Mechanism in [001] Direction. *Am. Mineral.* **2015**, *85*, 980–986. [CrossRef]
10. Hobbs, B.E. Recrystallization of single crystals of quartz. *Tectonophysics* **1968**, *6*, 353–401. [CrossRef]
11. Green, H.W. Metastable growth of coesite in highly strained quartz. *J. Geophys. Res.* **1972**, *77*, 2478–2482. [CrossRef]
12. Hirth, G.; Tullis, J. The brittle-plastic transition in experimentally deformed quartz aggregates. *J. Geophys. Res.* **1994**, *99*, 11731–11747. [CrossRef]

13. Richter, B. Stresses and pressures at the quartz-to-coesite phase transition in shear deformation experiments. *J. Geophys. Res.-Solid Earth* **2016**, *121*, 8015–8033. [CrossRef]
14. Pan, W.; Li, H.; Cui, C.; Jiang, J. In situ surface-enhanced Raman spectroscopy study of thiocyanate ions adsorbed on silver nanoparticles under high pressure. *Chem. Phys.* **2018**, *516*, 1–5.

minerals

MDPI

Article

Raman Spectroscopic Studies of Pyrite at High Pressure and High Temperature

Juan Chen [1], Heping Li [2,*], Yi Yuan [3], Mengxue Zhang [1], Shuhang Shuai [1] and Jingjing Wan [1]

1 School of Chemical Engineering, Guizhou Institute of Technology, Guiyang 550003, China; chenjuan0445@git.edu.cn (J.C.); 2018070102@stu.git.edu.cn (M.Z.); 2018070123@stu.git.edu.cn (S.S.); 2019070117@stu.git.edu.cn (J.W.)
2 Key Laboratory of High-Temperature and High-Pressure Study of the Earth's Interior, Institute of Geochemistry, Chinese Academy of Sciences, Guiyang 550081, China
3 Analysis and Testing Center, Guizhou Institute of Technology, Guiyang 550003, China; yuanyi@git.edu.cn
* Correspondence: liheping@vip.gyig.ac.cn

Abstract: Variations in the Raman spectra of pyrite were studied from 113 to 853 K at room pressure with a Linkam heating and freezing stage, and for 297–513 K and pressures up to 1.9 GPa with a hydrothermal diamond anvil cell. All observed frequencies decreased continuously with an increase in temperatures up to 653 K at ambient pressure. Hematite began to form at 653 K, all pyrite had transformed to hematite (H) at 688 K, and the hematite melted at 853 K. An increase in temperature at every initial pressure (group 1: 0.5 GPa, group 2: 1.1 GPa, group 3: 1.7 GPa, group 4: 1.9 GPa), showed no evidence for chemical reaction or pyrite decomposition. Two or three Raman modes were observed because of crystal orientation or temperature-induced fluorescence effects. The pressure groups showed a decreasing trend of frequency with gradual heating. The interaction of pressure and temperature led to a gradual decrease in Ag and Eg mode at a lower pressure (0.5 GPa and 1.1 GPa) than other pressure groups. Pressure and temperature effects are evident for groups 1 and 2; however, for groups 3 and 4, the temperature shows a larger effect than pressure and leads to a sharp decrease in Ag and Eg modes.

Keywords: pyrite; Raman spectroscopy; high pressure; high temperature

Citation: Chen, J.; Li, H.; Yuan, Y.; Zhang, M.; Shuai, S.; Wan, J. Raman Spectroscopic Studies of Pyrite at High Pressure and High Temperature. *Minerals* **2022**, *12*, 332. https://doi.org/10.3390/min12030332

Academic Editors: Lidong Dai, Haiying Hu, Jianjun Jiang and Jordi Ibanez-Insa

Received: 27 January 2022
Accepted: 3 March 2022
Published: 8 March 2022

1. Introduction

The most common sulfide mineral pyrite FeS_2, which is common in ore deposits and results in acid mine drainage because of its specific physical and chemical properties, has attracted the interest of many researchers in the high-pressure and high-temperature field.

As with NaCl, pyrite has a cubic structure, with a non-symmorphic space group $T_h{}^6$(Pa3) and four formula units per unit cell. The Fe atoms and center of the S_2 pairs (in the dumb-bell structure) occupy face-centered cubic lattice sites [1,2]. According to group (Pa3) theory, the irreducible representation of FeS_2 vibrations is expressed as Γ = Ag + Eg + 3Tg + 2Au + 6Tu, and the Ag + Eg + 3Tg represents five Raman active modes [1]. These expressions indicate that one mode is (Ag, S–S in phase stretch) symmetric and one is doubly degenerate (Eg, S_2 librational mode), and the third is triply degenerate (Tg, Tg(1) and Tg(3) are coupled librational and stretching modes with Tg(2) being the S–S out-of-phase stretching mode). When the Raman signals of pyrite were excited by the 514.5-nm line of an argon-ion laser, the assignment of all Raman modes included Eg: 344 cm^{-1}, Tg(1): 350 cm^{-1}, Tg(2): 377 cm^{-1}, Ag: 379 cm^{-1}, and Tg(3): 430 cm^{-1}, which is a typical structure of anisotropic pyrite, according to Kleppe and Jephcoat [2].

Ambient Raman research on pyrite has been undertaken since the 1970s [1,3–5]. It has been suggested that pyrite has type-I and type-II Raman frequency positions [5], and three or fewer modes have been observed [3–5], depending on the crystal orientation, sample purity, sample surface smoothness, and laser source. Vogt et al. [1] proposed the complete first-order Raman spectra of pyrite, for the first time, by theoretical interpretation of the phonon frequencies.

Thermogravimetric analysis (TGA) and differential thermal analysis (DTA) have been used extensively to research the thermodynamic properties of pyrite at a high temperature and room pressure [6–12]. Schoenlaub [6] suggested the decomposition of pyrite by TGA and DTA in different gases. They found that pyrite decomposes to sulfur vapor and pyrrhotite in neutral (N_2) or reducing gas (e.g., CO). It transforms to sulfur dioxide and ferrous sulfide between 718 K (445 °C) and 793 K (520 °C), and then sulfur dioxide and hematite above 893 K (620 °C) in oxygen, and shows a less pronounced reaction activity in air but has a greater activity at higher temperatures. Pyrite reacts with carbon dioxide slowly and transforms to magnetite. According to Jorgensen and Moyle [7], a periodic thermal instability exists when pyrite particles are oxidized in air at 723 K (450 °C); the TGA and DTA results show that the sample temperature was unstable in the first 40 min and the XRD result demonstrated the presence of hematite and pyrite, with temperatures ranging between 643 K (370 °C) and 753 K (480 °C). Music et al. [8] studied the thermal decomposition of pyrite by XRD and ^{57}Fe Mossbauer spectroscopy in air and found that hematite, maghemite, and pyrrhotite were the thermal decomposition products of natural pyrite; only hematite was present above 673 K (400 °C). Fegley et al. [10] studied pyrite decomposition kinetics in pure CO_2 and impure CO_2 gas and found that pyrite transforms to pyrrhotite (Fe_7S_8) or more Fe-rich pyrrhotites, to magnetite (Fe_3O_4) by oxidation, to maghemite (γ-Fe_2O_3), and finally to hematite (α-Fe_2O_3). Li and Zhang [13] proposed temperature dependence susceptibilities of pyrite in argon and air and found that pyrite transforms to magnetic minerals (magnetite and pyrrhotite) when heated to 653 K (380 °C).

High-pressure studies of pyrite are also interesting [2,14–17]. Pyrite is structurally stable to 55 GPa at static pressure and to 320 GPa in shock wave experiments. Cervantes et al. [15] found that the band gap of FeS_2 decreased linearly with pressure at -1.13×10^{-2} eV/GPa to 28 GPa. According to the linear trend, FeS_2 is expected to metallize at a pressure of 80 (± 8) GPa. Kleppe and Jephcoat [2] described the Raman shift variation in pyrite under hydrostatic and non-hydrostatic conditions to 55 GPa at room temperature. Liu et al. [17] conducted research on the electrical conductivity at high temperature and pressure. They found that the electrical conductivities of natural pyrite increased with pressure, and the increasing trend strengthened after ~13 GPa, which correlates to the energy level of the trace element Co in pyrite.

Simultaneous high-pressure and high-temperature studies of pyrite have been undertaken [16,17]. Liu et al. [17] described the variation in the electrical conductivity up to 573 K and from 1 atm to 20.9 GPa. They found that the electrical conductivity increased with temperature at a constant pressure, and so does the pressure, and no evidence of chemical reaction or structural phase transition of pyrite was detected by Raman spectroscopy. Yuan and Zheng [16] found that the positive dependence of Ag and Eg modes on pressure indicates a stress-induced contraction of S–S and Fe–S bonds, whereas the negative dependence on temperature shows their temperature-induced expansion. The thermal expansion coefficient was derived from the Raman spectra of pyrite up to 2.1 GPa and 675 K. A 20 wt.% NaCl solution was used as the pressure-transmitting medium by Yuan and Zheng [16].

A study of pyrite at 113–873 K and atmospheric pressure by Raman spectra has not been developed adequately. Therefore, this study aims to establish the Raman spectroscopic characteristics of FeS_2 pyrite under the abovementioned conditions. The Raman spectra of pyrite from 0.1 MPa to 2 GPa and up to 553 K under hydrostatic conditions is also inadequate, and further detailed study is necessary to confirm the dominating factor of pressure and/or temperature on Raman shifts.

2. Materials and Methods

Pyrite samples were collected from Longsheng Ethnic Minorities Autonomous County in Guangxi. The mineral phase of the pyrite was confirmed by an an X'Pert Pro X-ray diffractometer (Phillips Company, Amsterdam, The Netherlands), with a purity above 99.0%, and the chemical composition of major pyrite (Table 1) was determined by a EPMA-1600 electron probe microanalyzer (Shimadzu, Kyoto, Japan) in the State Key Laboratory of Ore Deposit Geochemistry, Institute of Geochemistry, Chinese Academy of Sciences. Some samples were cut to 10 mm × 10 mm × 0.2 mm and polished on both sides for high- and low-temperature experiments on a heating and freezing stage. Other samples were ground to pass through an 80 mesh sieve for simultaneous high-pressure and high-temperature experiments in a hydrothermal diamond anvil cell (HDAC).

Table 1. Chemical composition of major pyrite analyzed with EPMA (wt.%).

Pyrite	S (%)	Fe (%)	Hg (%)	Mo (%)	Au (%)	Co (%)	Pb (%)	Zn (%)	Total (%)
No.1	52.43	46.37	1.08	0.47	0.00	0.10	0.11	0.06	100.62
No.2	52.77	46.49	0.56	0.57	0.20	0.11	0.10	0.06	100.87
No.3	52.64	45.97	0.66	0.51	0.34	0.09	0.07	0.00	100.28
No.4	52.54	46.18	0.00	0.50	0.05	0.09	0.12	0.18	99.67
average	52.60	46.25	0.57	0.51	0.15	0.10	0.10	0.07	100.36

The spectra were excited by the 514.5 nm line of a Spectra-Physics 2017 laser (Newport Corporation, Irvine, CA, USA) and collected through a Renishaw InVia Raman spectrometer (Renishaw, London, England) in the Key Laboratory of High-Temperature and High-Pressure Study of the Earth's Interior, Institute of Geochemistry, Chinese Academy of Sciences, the spectrometer resolution was ±1 cm^{-1}.

2.1. High- and Low-Temperature Experiments

The sample was placed directly on a quartz window in a Linkam THMSG600 heating and freezing stage (Linkam Scientific Instruments, Tadworth, UK). Liquid nitrogen purging was used to protect the stage and provide low temperatures of 113 K–273 K. A platinum resistor sensor was used as a heater to provide high temperatures at 273 K–873 K. The temperature was monitored by a platinum (10%) rhodium/platinum Type-S thermocouple that was attached to the heater. The temperature resolution was 0.01 K, and the temperature was stable to ±0.1 K.

In the heating and cooling studies at room pressure, the working parameters of the Raman spectrometer included the exposure time of 10 s, a 2-times accumulations, laser power of 20 MW, spectrum range of 100–1500 cm^{-1}, and objective of SLM PLAN 20×.

2.2. Simultaneous High-Pressure and High-Temperature Experiments

Heating in the HDAC can be achieved with small resistance furnaces around the diamond anvils. The HDAC is similar to a Bassett-type [18] externally heated diamond anvil cell. A 0.3-mm-thick stainless-steel sheet (model T301) was used as a gasket and the sample chamber was 0.3–0.4 mm in diameter and placed between the two 0.8-mm culets as a sandwich. A methanol–ethanol ratio of 4:1 was loaded into the HDAC chamber with a single quartz crystal. The experimental pressure was calculated from Equation (1) according to the Raman shift of the 464 cm^{-1} peak of quartz [19], the uncertainty of the pressure was ±50 MPa at 23 °C. The difference in frequency of the 464 cm^{-1} line at the experimental temperature T and the frequency at 23 °C is described in the α-quartz stability field by Equation (2), where $-196 \le T\,(°C) \le 560$.

$$P(\text{MPa}) = 0.36079 \times [(\Delta\nu_P)_{464}]^2 + 110.86 \times (\Delta\nu_P)_{464} \quad (1)$$

$$(\Delta\nu_T)_{464,\ P=0.1\text{MPa}}(\text{cm}^{-1}) = 2.50136 \times 10^{-11} \times T^4 + 1.46454 \times 10^{-8} \times T^3 - 1.801 \times 10^{-5} \times T^2 - 0.01216 \times T + 0.29 \quad (2)$$

The sample was heated by small furnaces made of $Ni_{90}Cr_{10}$ alloy wires and the temperature was measured by a $Ni_{90}Cr_{10}$–$Ni_{95}Si_5$ thermocouple, which was calibrated by referring to the melting points of C_6H_5NH ($COCH_3$) (acetanilide, 388 K (115 °C)) and $C_{20}H_{14}O_4$ (phenolphthalein, 535.6 K (262.6 °C)) at atmospheric pressure before the experiment. The uncertainty of the temperature was ±2.5 K in the range of 273 K–553 K (0–280 °C).

In the simultaneous high-pressure and high-temperature study, the following Raman spectrometer parameters were used: 30 s exposure time, 2-times accumulation, 80 MW laser power, 100–1000 cm^{-1} spectral range, and SLM PLAN 20× of objective.

3. Results and Discussion

3.1. High- and Low-Temperature Study of Pyrite at Ambient Pressure

The Raman peak assignment of pyrite is given in Table 2 [2]. Only three Raman active modes of pyrite were observed (Eg, Ag, Tg(3)), which agrees with most studies. Sometimes only two modes, Eg and Ag, were observed, which depended on crystal orientation perpendicular to the laser emission.

Table 2. Ambient Raman bands (cm^{-1}) of pyrite at 298 K.

ν_0 (cm^{-1}) (This Study)	ν_0 (cm^{-1}) (Kleppe and Jephcoat [2])	Comments	Symmetries and Assignments
342	344		Eg, S_2 libration
348	350	Weak and need peak separation	Tg(1), coupled libration and stretch
378	379	377 cm^{-1} is covered by 379 cm^{-1}	Ag, S–S in phase stretching Tg(2), S–S out-of-phase stretching
429	430	Appear in certain orientations and lower temperature	Tg(3), coupled libration and stretch

A weakened Raman peak signal that was caused by the fluorescence effects of high temperature was observed, but limited effects resulted for the low-temperature condition, which is supported by the low-temperature study of pyrite. The Raman frequencies for the Eg, Ag, and Tg(3) bands of pyrite decreased with an increase in temperature in Figures 1 and 2, which is consistent with the results of Yuan and Zheng [16].

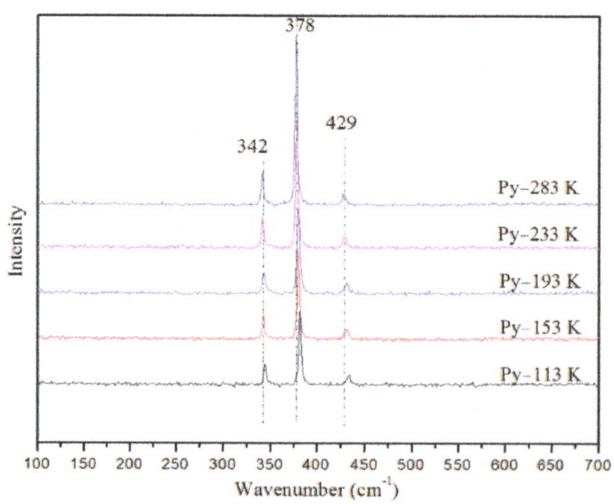

Figure 1. Raman shift of pyrite during cooling (113–283 K).

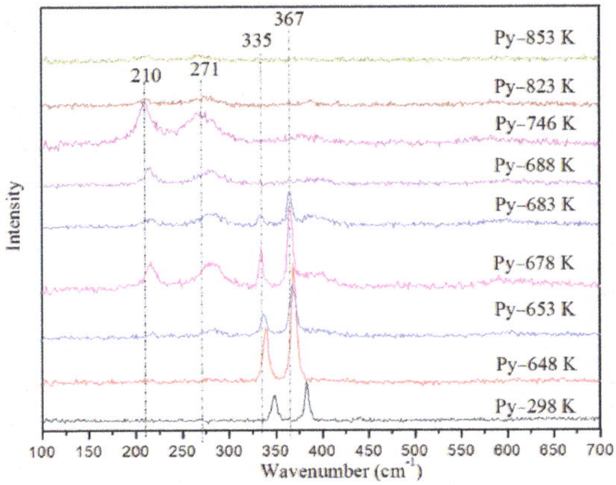

Figure 2. Raman shift of pyrite during heating (298–853 K).

The Eg, Ag, and Tg(3) Raman wavenumbers are shown in Figure 3. The quadratic temperature dependence is within the experimental error. The parameters that were derived by quadratic and linear regressions of all modes are shown in Tables 3 and 4. The temperature dependence of the Ag mode is larger than that in Eg mode (-0.0366 cm^{-1}/K compared with -0.0245 cm^{-1}/K), which led to obvious separation between them with temperature, because the Ag (in the phase S–S stretching frequency) mode was determined mainly by the S–S force constant [20] and the Eg (librational frequency) was largely governed by the Fe–S force constant. Temperature induces faster molecule vibrations and a decreasing S–S and Fe–S bond energy. The Tg(3) mode disappeared above 303 K because of the weakened Raman signal at the high temperature (Figure 3). It was not known whether a chemical reaction or phase transition resulted because no new frequencies appeared. At 653 K, two new peaks were observed: (H-(Eg(2) + Eg(3)) represents the mixture modes of Eg(2) and Eg(3) of hematite, and H-A1g(1) represents the A1g(1) mode of hematite. At 688 K, the Ag and Eg modes of pyrite disappeared, which left two new peaks until 853 K. It is speculated that the new frequencies represent hematite from the Raman library [21]. To confirm the new phase, we conducted another experiment, where the temperature was reduced gradually to 293 K after it had been increased to 703 K. Figure 4 compares the pyrite product to a previous study on hematite [22], and the two spectra appeared consistent, except for the broader peak and high background in this study.

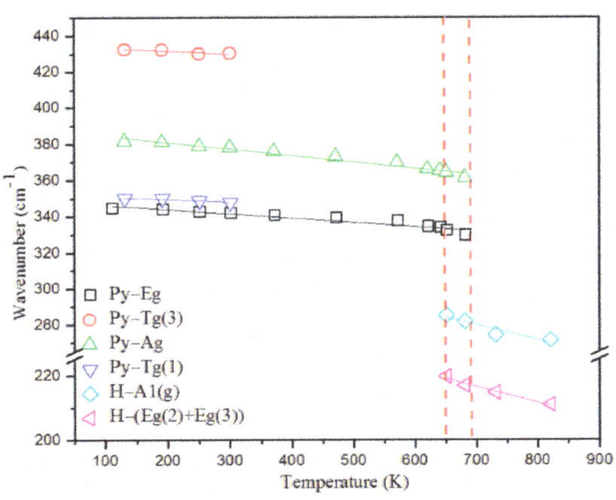

Figure 3. Wavenumber of six peak variations with increasing temperature.

Table 3. Variable modes of Raman bands (Δν) of pyrite with increasing temperature according to νi = aT^2 + bT + c. νi and Δν are in cm^{-1}, T in K, a and b have the corresponding units, c is a constant, and R^2 is the correlation coefficient.

ν_i (cm^{-1})	Δν	a × 10^5	b × 10^2	c	R^2
Eg: 329.94	−14.432	−4	1.09	342.91	0.9780
Tg(1): 347.87	−2.393	−9	2.72	347.97	0.8504
Ag: 361.25	−20.302	−5	0.78	381.01	0.9907
Tg(3): 430.28	−2.405	7	−4.66	437.99	0.8599
H-(Eg(2) + Eg(3)): 274.19	−11.201	−4	1.09	342.91	0.9780
H-A1g(1): 211.04	−8.675	20	−32.84	353.89	0.9709

Table 4. Variable modes of Raman bands (Δν) of pyrite with increasing temperature according to νi = aT + c. νi and Δν are in cm^{-1}, T in K, a in cm^{-1}/K, c is a constant, and R^2 is the correlation coefficient.

ν_i (cm^{-1})	Δν	a (cm^{-1}/K)	c	R^2
Eg: 329.94	−14.432	−0.0245	348.49	0.9394
Tg(1): 347.87	−2.393	−0.0129	352.05	0.7549
Ag: 361.25	−20.301	−0.0366	388.13	0.9654
Tg(3): 430.28	−2.405	−0.0181	435.09	0.8289
H-(Eg(2) + Eg(3)): 274.19	−11.201	−0.0788	336.08	0.8123
H-A1g(1): 211.04	−8.675	−0.0505	251.91	0.9426

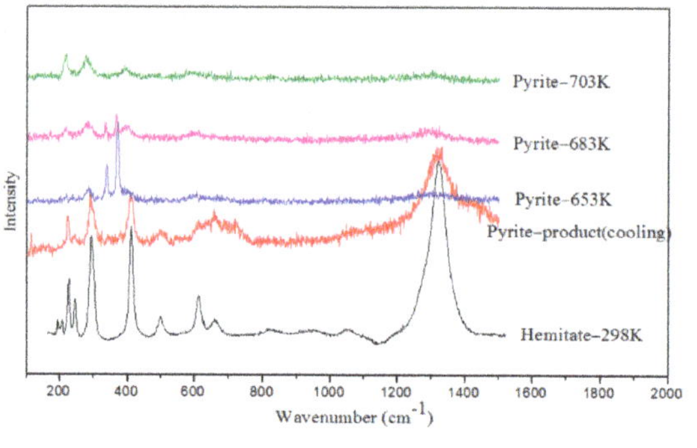

Figure 4. Raman spectra of pyrite and hematite from bottom to top: hematite at 298 K [22], increased pyrite temperature to 703 K and cooling to 298 K, pyrite at 683 K, pyrite at 703 K.

Pyrite began to transform to hematite from oxidation at 653 K. At ambient conditions, the hematite phonon frequencies were A1g(1): 224 cm^{-1}, Eg(1): 243 cm^{-1}, Eg(2): 290 cm^{-1}, Eg(3): 297 cm^{-1}, Eg(4): 408 cm^{-1}, A1g(2): 496 cm^{-1}, and Eg(5): 609 cm^{-1} (Shim and Duffy, 2001). Two broad and strong peaks were visible: 285 cm^{-1} represents a mixture of Eg(2) and Eg(3), and 219 cm^{-1} represents A1g(1) in the high-temperature study. Figure 2 shows the phase transformation from pyrite to hematite according to the Raman shift. Pyrite and hematite coexist at 653 K. At 688 K, the Eg and Ag modes of pyrite disappeared. The poor crystal of hematite resulted in melting at 823 K and the hematite melted at 853 K, where no signal was observed because of the temperature-induced fluorescence effect and weak focus of melts.

The high-temperature Raman study of pyrite by Linkam heating and cooling indicates that pyrite decomposed to hematite between 653 K and 688 K. The results from this study are distinct from a previous study [6] but like those of Jorgensen and Moyle [7], and Music et al. [8] as shown in Figure 5. Jorgensen and Moyle [7] reported the presence of hematite between 643 K (370 °C) and 753 K (480 °C); Similarly, hematite was present at

673 K (400 °C) according to Music et al. [8]. The difference in temperatures for hematite's presence between this study and the two abovementioned experiments could have resulted from an inconsistent heating rate. This study may provide more accurate data by virtue of the advanced Linkam heating and freezing stage technology. Further studies will be required to check the effect of heating rate on the decomposition temperature.

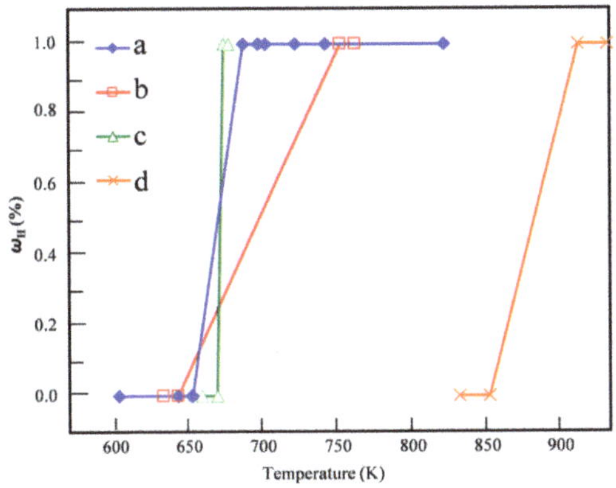

Figure 5. Hematite content produced as a product of pyrite with increasing temperature, ω_H represents the mass fraction of hematite. (**a**) Present work (solid diamond). (**b**) Jorgensen and Moyle [7] (open square). (**c**) Music et al. [8] (open triangle). (**d**) Schoenlaub [6] (red cross).

The full width at half maximum (FWHM) of six bands displays the temperature dependence (Figure 6). The FWHM of all pyrite modes broadens with an increase in temperature, but the hematite modes (H-A1g(1) and H-(Eg(2) + Eg(3)) increase sharply with an increase in temperature because the hematite that is produced by pyrite at a high temperature has a poor crystal structure and rough surface.

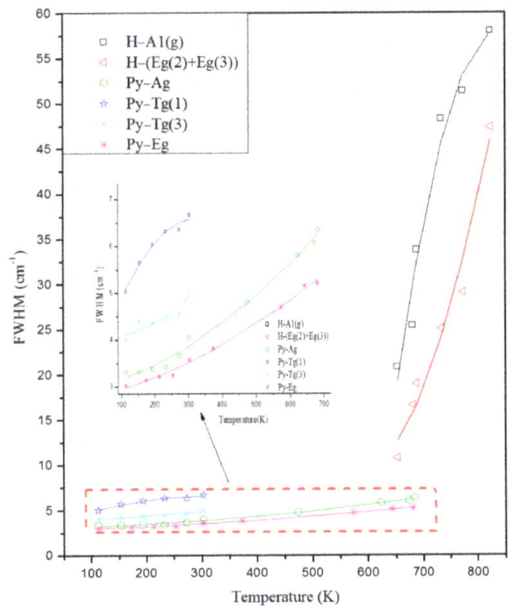

Figure 6. FWHM of six bands' variation with increasing temperature.

3.2. Simultaneous High-Temperature and High-Pressure Study of Pyrite

Water reacts with pyrite as a function of temperature, pressure, and time. To avoid the reaction between water and pyrite, methanol–ethanol (4:1) was used as a pressure medium instead of a 20 wt.% NaCl solution [16] to provide hydrostatic pressure. The synthetic quartz Raman band [19] was measured as a pressure calibrant in this study. High-temperature experiments were conducted at four different initial pressures (group 1: 0.5 GPa, group 2: 1.1 GPa, group 3: 1.7 GPa, group 4: 1.9 GPa). A common feature was observed in Figure 7, i.e., pressure increases slowly to 453 K. The cell's mechanical properties were influenced above 453 K and resulted in an induced decrease in pressure. Test replication would help to improve the data, so that pressure variation that is associated with this condition is negligible.

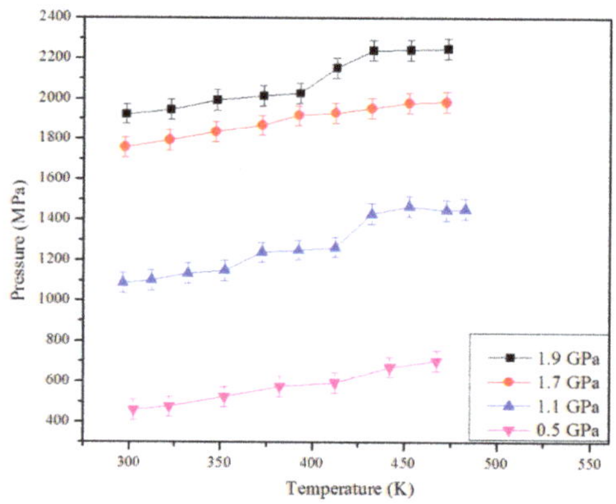

Figure 7. Various initial pressures, pyrite heating in HDAC from 298 K–493 K, in situ pressures (0.5, 1.1, 1.7, 1.9 GPa) at every temperature point. Error bars show pressure uncertainty is ±50 MPa.

Both the temperature and pressure effect have an impact (Figure 8), and the Ag and Eg mode shifts to being faster for groups 4 and 3 than for groups 1 and 2. The positive pressure and negative temperature dependence of the two modes suggests that the pressure and temperature effects are identically strong at a lower pressure (see groups 1 and 2), and the temperature dependence is more obvious at a higher pressure (see groups 3 and 4).

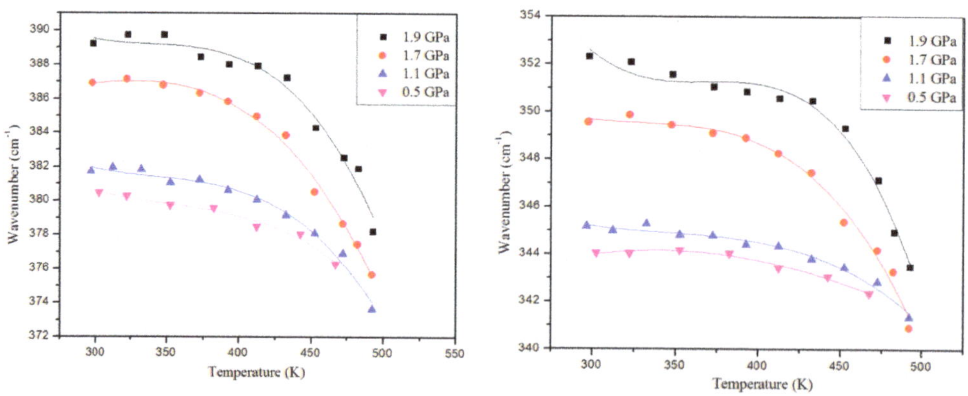

(a) (b)

Figure 8. (**a**) Raman shift of A(g) mode heating at different initial pressures: 0.5, 1.1, 1.7, 1.9 GPa; (**b**) Raman shift of E(g) mode heating at different initial pressures: 0.5,1.1, 1.7, 1.9 GPa.

4. Conclusions

All observed pyrite frequencies decreased continuously with an increase in temperature to 653 K at room pressure. Pyrite transformed to hematite at 653 K, had fully changed to hematite at 688 K, and the hematite melted at 853 K. At the four initial pressures (0.5, 1.1, 1.7, 1.9 GPa) and with an increasing temperature, there was no evidence for chemical reaction or decomposition. Every group demonstrated a frequency decline with gradual heating. The interaction of pressure and temperature led to a gentle decrease in Ag and Eg modes at a lower pressure (0.5 and 1.1 GPa) than other groups. The positive pressure and negative temperature dependence of the two modes suggests that the pressure and temperature effect are identically strong at a lower pressure (0.5 and 1.1 GPa), and the temperature dependence is more prominent at a higher pressure (1.7 and 1.9 GPa).

Author Contributions: Methodology, J.C. and H.L.; software, Y.Y.; resources and data curation, M.Z., S.S. and J.W.; writing—original draft preparation, J.C.; writing—review and editing, H.L.; funding acquisition, J.C. and H.L. All authors have read and agreed to the published version of the manuscript.

Funding: This research is supported by the fund from the Science and Technology Agency of Guizhou province, China ([2020]1Y170 and [2019]2830).

Data Availability Statement: Not applicable.

Acknowledgments: The authors thank Terry Mernagh for grammar modification.

Conflicts of Interest: The authors declare no conflict of interest.

References

1. Vogt, H.; Chattopadhyay, T.; Stolz, H.J. Complete first-order Raman spectra of the pyrite structure compounds FeS_2, MnS_2 and SiP_2. *J. Phys. Chem. Solids* **1983**, *44*, 869–873. [CrossRef]
2. Kleppe, A.K.; Jephcoat, A.P. High-pressure Raman spectroscopic studies of FeS_2 pyrite. *Mineral. Mag.* **2004**, *68*, 433–441. [CrossRef]
3. Ushioda, S. Raman scattering from phonons in iron pyrite (FeS_2). *Solid State Commun.* **1972**, *10*, 307–310. [CrossRef]
4. Anastassakis, E.; Perry, C.H. Light scattering and IR measurements in XS2 pryite-type compounds. *J. Chem. Phys.* **1976**, *64*, 3604–3609. [CrossRef]
5. Mernagh, T.P.; Trudu, A.G. A laser Raman microprobe study of some geologically important sulphide minerals. *Chem. Geol.* **1993**, *103*, 113–127. [CrossRef]
6. Schoenlaub, R.A. Oxidation of pyrite. *J. Am. Ceram. Soc.* **1969**, *52*, 40–43. [CrossRef]
7. Jorgensen, F.; Moyle, F.J. Periodic thermal instability during the isothermal oxidation of pyrite. *Metall. Mater. Trans. B* **1981**, *12*, 769–770. [CrossRef]
8. Music, S.; Popović, S.; Ristić, M. Thermal decomposition of pyrite. *J. Radioanal. Nucl. Chem.* **1992**, *162*, 217–226. [CrossRef]
9. Inge, I.M.; Jan, Y.; Heidi, V.; Dirk, V.F.; Jules, M.; Lucien, C.V.P.; Grazyna, G.; Piotr, W. Study of coal-derived pyrite and its conversion products using atmospheric pressure temperature-programmed reduction (AP-TP). *Energy Fuels* **1995**, *9*, 950–955.
10. Fegley, B., Jr.; Lodders, K.; Treiman, A.H.; Klingelhöfer, G. The rate of pyrite decomposition on the surface of Venus. *Icarus* **1995**, *115*, 159–180. [CrossRef]
11. Corradi, A.B.; Leonelli, C.; Manfredini, T.; Romagnoli, M. Quantitative determination of pyrite in, ceramic clay raw materials by DTA. *Thermochim. Acta* **1996**, *287*, 101–109. [CrossRef]
12. Yi, P.; Yu, Q.; Zong, H. Thermodynamic analysis for chemical desulfurization of pyrite in coal. *Coal Convers.* **1999**, *22*, 47–52.
13. Li, H.; Zhang, S. Detection of mineralogical changes in pyrite using measurements of temperature- dependence susceptibilities. *Chin. J. Geophys.* **2005**, *48*, 1384–1391. [CrossRef]
14. Merkel, S.; Jephcoat, A.P.; Shu, J.; Mao, H.; Gillet, P.; Hemley, R.J. Equation of state, elasticity, and shear strength of pyrite under high pressure. *Phys. Chem. Miner.* **2002**, *29*, 1–9. [CrossRef]
15. Cervantes, P.; Slanic, Z.; Bridges, F.; Knittle, E.; Williams, Q. The band gap and electrical resistivity of FeS_2-pyrite at high pressures. *J. Phys. Chem. Solids* **2002**, *63*, 1927–1933. [CrossRef]
16. Yuan, X.; Zheng, H. In situ Raman spectroscopic studies of FeS_2 pyrite up to 675 K and 2100 MPa using a hydrothermal diamond anvil cell. *Mineral. Mag.* **2015**, *79*, 1–10. [CrossRef]
17. Liu, K.; Dai, L.; Li, H.; Hu, H.; Wu, L.; Zhuang, Y.; Pu, C.; Yang, L. Migration of impurity level reflected in the electrical conductivity variation for natural pyrite at high temperature and high pressure. *Phys. Chem. Miner.* **2018**, *45*, 85–92. [CrossRef]
18. Bassett, W.A. High pressure-temperature aqueous systems in the hydrothermal diamond anvil cell (HDAC). *Eur. J. Mineral.* **2003**, *15*, 773–780. [CrossRef]
19. Schmidt, C.; Ziemann, M.A. In-situ Raman spectroscopy of quartz: A pressure sensor for hydrothermal diamond-anvil cell experiments at elevated temperatures. *Am. Mineral.* **2000**, *85*, 1725–1734. [CrossRef]

20. Lutz, H.D.; Zwinscher, J. Lattice dynamics of pyrite FeS_2 -polarizable-ion model. *Phys. Chem. Miner.* **1996**, *23*, 497–502. [CrossRef]
21. Database of Raman Spectra. X-ray Diffraction and Chemistry Data for Minerals. Available online: https://rruff.info/hematite/display=default/R050300 (accessed on 10 January 2022).
22. Shim, S.; Duffy, T. Raman spectroscopy of Fe_2O_3 to 62 GPa. *Am. Mineral.* **2002**, *87*, 318–326. [CrossRef]

minerals

MDPI

Article

Pressure-Induced Reverse Structural Transition of Calcite at Temperatures up to 873 K and Pressures up to 19.7 GPa

Xinyu Zhang [1,2], Lidong Dai [1,*], Haiying Hu [1,*] and Chuang Li [1,2]

1 Key Laboratory of High-Temperature and High-Pressure Study of the Earth's Interior, Institute of Geochemistry, Chinese Academy of Sciences, Guiyang 550081, China
2 University of Chinese Academy of Sciences, Beijing 100049, China
* Correspondence: dailidong@vip.gyig.ac.cn (L.D.); huhaiying@vip.gyig.ac.cn (H.H.)

Abstract: In situ Raman scattering and electrical conductivity experiments have been performed to investigate the structural phase transitions of calcite during the compressed and decompressed processes in a diamond anvil cell at temperatures of 298–873 K and pressures up to 19.7 GPa. Upon compression, calcite ($CaCO_3$-I phase) underwent three structural phase transitions from $CaCO_3$-I to $CaCO_3$-II phases at 1.6 GPa, from $CaCO_3$-II to $CaCO_3$-III phases at 2.2 GPa, and from $CaCO_3$-III to $CaCO_3$-VI phases at 16.8 GPa under room temperature conditions, which were evidenced by the evolution of Raman peaks, as well as the discontinuities in the pressure-dependent Raman shifts and electrical conductivity. Upon decompression, the structural phase transitions from $CaCO_3$-VI to $CaCO_3$-III to $CaCO_3$-II to $CaCO_3$-I phases took place at the respective pressures of 5.4, 1.5, and 0.4 GPa, indicating the reversibility of calcite. Furthermore, an obvious ~11 GPa of pressure hysteresis was detected in the $CaCO_3$-VI to $CaCO_3$-III phase transition, whereas other reverse phase transition pressures were very close to those of compressed results. At three given representative pressure conditions (i.e., 10.5, 12.5, and 13.8 GPa), a series of electrical conductivity experiments were performed at temperature ranges of 323–873 K to explore the temperature-dependent relation of $CaCO_3$-III to $CaCO_3$-VI structural phase transition. With increasing pressure, the transition temperature between $CaCO_3$-III and $CaCO_3$-VI phases gradually decreases, which reveals an obviously negative temperature-pressure relation, i.e., P (GPa) = 19.219 (±1.105) − 0.011 (±0.002) T (K). Our acquired phase diagram of calcite can be employed to understand the high-pressure structural transitions and phase stability for carbonate minerals along various subducting slabs in the deep Earth's interior.

Keywords: calcite; phase diagram; electrical conductivity; Raman spectroscopy; structural phase transition; high temperature; high pressure

Citation: Zhang, X.; Dai, L.; Hu, H.; Li, C. Pressure-Induced Reverse Structural Transition of Calcite at Temperatures up to 873 K and Pressures up to 19.7 GPa. *Minerals* **2023**, *13*, 188. https://doi.org/10.3390/min13020188

Academic Editor: Jordi Ibanez-Insa

Received: 2 December 2022
Revised: 24 January 2023
Accepted: 26 January 2023
Published: 27 January 2023

1. Introduction

In nature, calcium-carbonate minerals crystallize in three different structural forms: rhombohedral calcite, orthorhombic aragonite and hexagonal vaterite. Among those of calcium-carbonate minerals, calcite is an important and common rock-forming mineral in the Earth's crust, and is considered as a crucial carrier of carbon. Calcite plays a key role in the deep carbon cycle of the Earth's interior, which can be transported into the interior of the Earth through the subduction of oceanic crust [1–8]. Thus, it is vital to investigate the structural transition and phase stability for calcite under high-temperature and high-pressure conditions.

A great number of Raman scattering investigations on the structural transition and phase stability of calcite have been performed under conditions of room temperature and high pressure [9–12]. For most of these previously acquired results, three main high-pressure polymorphs of calcite ($CaCO_3$-I phase) were disclosed, i.e., monoclinic calcite ($CaCO_3$-II phase), layered triclinic calcite ($CaCO_3$-III phase), and non-layered triclinic calcite ($CaCO_3$-VI phase). The structural phase transitions for these of $CaCO_3$-I to $CaCO_3$-II phases and $CaCO_3$-II to $CaCO_3$-III phases have been well documented at lower pressure points of 1.6 GPa and 2.0 GPa by means of Raman scattering spectroscopy in a diamond anvil cell at room temperature [9,10]. However, the high-pressure crystalline structure of $CaCO_3$-III phase and its corresponding transformation pressure remains controversial in all of the listed previously reported results. For example, Suito et al. explored the structural phase transition of calcite through Raman scattering measurement in combination with synchrotron X-ray experiments at room temperature and pressures up to 10.0 GPa [11]. Their results showed that $CaCO_3$-I to $CaCO_3$-II and $CaCO_3$-II to $CaCO_3$-III phase transitions appear at the respective pressures of 1.7 GPa and 2.1 GPa. The authors further increased pressure to 10.0 GPa, and totally new Raman peaks exhibiting the discrepancy with the $CaCO_3$-III phase were observed, which was possibly related to a pressure-induced structural phase transition from $CaCO_3$-III phase to aragonite. Subsequently, Liu et al. investigated the high-pressure behavior of calcite up to 75.0 GPa by virtue of Raman spectroscopy and theoretical calculations [12]. They revealed that the $CaCO_3$-I to $CaCO_3$-III phase transition takes place at 2.0 GPa, and further, the $CaCO_3$-III phase transforms into the $CaCO_3$-VI phase instead of aragonite at 16.0 GPa, which is inconsistent with the high-pressure Raman scattering results to be reported by Suito et al. Therefore, further Raman spectroscopic investigation on calcite is indispensable to explore its vibrational properties with increasing pressure at room temperature.

On the other hand, high-pressure electrical conductivity measurement is also an effective experimental means of extrapolating structural transition and phase stability for carbonate minerals at high temperatures [13–17]. Bagdassarov and Slutskii conducted the electrical conductivity measurement of calcite under temperature ranges from 600 K to 1200 K and pressures of 0.5–2.5 GPa in a piston cylinder apparatus to establish the phase boundaries from calcite to aragonite, from $CaCO_3$-I to $CaCO_3$-IV phases and from $CaCO_3$-IV to $CaCO_3$-V phases [13]. However, owing to the uppermost limited pressure range in the piston cylinder apparatus, their maximum experimental pressure is only up to 2.5 GPa under simultaneous high-temperature and high-pressure conditions. Therefore, it is important to investigate the structural transition and phase stability for calcite by means of electrical conductivity measurement under broader temperature and pressure ranges.

In the present work, we have investigated the structural transition and phase stability for calcite under conditions of 298–873 K and pressures up to 19.7 GPa in a diamond anvil cell by in situ Raman spectroscopy and electrical conductivity measurements. Three high-pressure structural transformations from $CaCO_3$-I to $CaCO_3$-II to $CaCO_3$-III to $CaCO_3$-VI phases were revealed at the respective pressures of 1.6, 2.2, and 16.8 GPa under atmospheric temperature. At three fixed pressure points (i.e., 10.5, 12.5, and 13.8 GPa), the phase transition temperatures between $CaCO_3$-III and $CaCO_3$-VI phases were individually determined. Furthermore, a pressure-temperature phase diagram of calcite was well established at the temperatures of 298–773 K, pressures up to 16.0 GPa and depth up to 480 km, and further, its potential geophysical implications were discussed in detail.

2. Materials and Methods

2.1. Sample Preparation and Characterization

Naturally transparent single-crystal calcite (Spec: 3.0 cm $\times$ 3.0 cm $\times$ 2.0 cm) was gathered from Nanyang ore district in Qingyang county, Anhui province, China. We crushed and ground the initial single-crystal calcite into powder in an agate mortar. The crystalline structure of the starting sample was characterized by X-ray diffraction (XRD) experiments operating with a PIXcel3D area detector (X-ray wavelength: 0.154056 nm; collection time: 4 min; beam size: 0.5 mm; working voltage: 2.2 kV and resolution: 0.028°) in the State Key Laboratory of Ore Deposit Geochemistry, Institute of Geochemistry, Chinese Academy of Sciences. As shown in Figure 1, the obtained XRD pattern can be well indexed with a trigonal crystalline system of the space group of D_{3d}^5: $R\overline{3}c$. The GSAS software package was employed to fit and refine the obtained XRD data [18]. The lattice parameters of calcite were determined as $a = b = 4.988 \pm 0.003$ Å, $c = 17.052 \pm 0.002$ Å; $\alpha = \beta = 90°$,

$\gamma = 120°$; and $V = 367.48 \pm 0.56$ Å^3 (JCPDS NO. 83-0578), which agrees well with previously reported results [19,20].

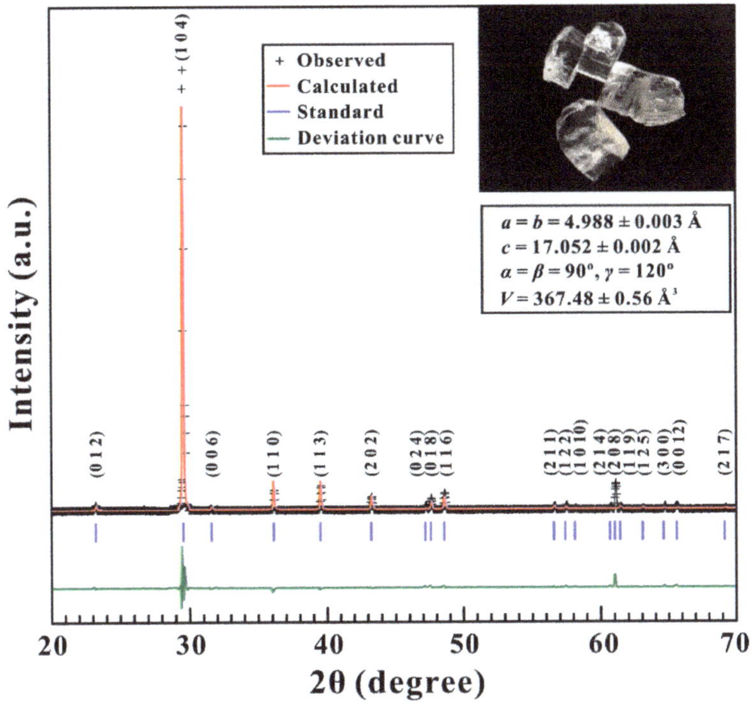

Figure 1. Powder X-ray diffraction (XRD) pattern and structural refinement for calcite at ambient conditions. The red solid line and the black crosses represent the calculated and observed results, respectively. The vertical blue bars stand for the standard peak positions of calcite. The green solid line denotes the deviation curve of measurement. Inset: the optical image of calcite.

2.2. High-Pressure Raman Spectroscopy Measurements

High-pressure Raman scattering experiments of calcite were carried out using a piston-cylinder diamond anvil cell (DAC) with a 300-μm diameter anvil culet. A piece of custom-built T-301 stainless steel (Spec: the area of 5×5 mm^2 and its corresponding thickness of 250 μm) was utilized as the gasket and then was pre-indented into an approximate 50-μm thickness. In the center of the pre-indented gasket, one new 100-μm hole was drilled as the sample chamber. Subsequently, calcite powder samples together with several tiny grains of ruby were sealed into the sample chamber. The pressure inside the sample chamber was calibrated on the basis of the wavenumber shift of the fluorescence peak of the trivalent chromic ion from ruby [21]. The high-pressure Raman scattering experiments of calcite were conducted in the backscattering geometry by means of a confocal Renishaw 2000 Raman spectrometer with an Olympus charge-coupled device (CCD) camera. The 514.5 nm wavenumber of Argon ion laser was applied as the excitation laser with the magnitude of powers of 1 mW and 0.5–50 μW for high-pressure Raman measurements of calcite and each corresponding pressure calibration of ruby, respectively. The Raman spectra of calcite were collected within the wavenumber range of 100–1200 cm^{-1} with a resolution of 0.1 cm^{-1} at each pre-designed pressure. The acquisition time was set as 240 s for each individual Raman spectroscopy measurement.

2.3. High-Pressure Electrical Conductivity Measurements

High-temperature and high-pressure electrical conductivity measurements of calcite were performed using a well-designed four column-type DAC with a 300-μm diameter taking advantage of its large lateral space, which has been widely applied to our previous electrical transport experiments [22–34]. A mixture of cubic boron nitride (c-BN) and epoxy resin powders was utilized to constitute an insulating sample chamber. Then, a new central hole with the 120-μm diameter was made by a laser drilling machine. Plate electrodes of platinum with 4-μm thickness were integrated separately onto the upper and lower counterparts of the sample chamber. Prior to the electrical conductivity measurement, experimental samples were heated to 373 K in a vacuum oven for 12 h to eliminate the absorbed water. The alternating current (AC) impedance spectroscopy of calcite was measured using the combination interface of Solartron-1260 and Solartron-1296 impedance analyzers within the frequency range of 10^{-1}–10^{7} Hz at a fixed signal voltage of 1.0 V. High-temperature conditions were provided by two external resistance heating furnaces which were tightly twisted around the tungsten carbide bases. A *k*–type thermocouple (NiCr–NiAl) with temperature error of ±5 K was attached to the side face of a diamond to monitor the temperature of the sample chamber. For high-temperature and high-pressure electrical conductivity measurements, we firstly compressed the sample to pre-determined pressure values, and gradually heated the sample chamber from 298 K to 873 K with a temperature interval of 50 K. Then, the temperature was sustained for at least ten minutes to achieve the thermal transference equilibrium. In order to protect the sample assemblage, a cooling system equipped with circulating water was used under high-temperature conditions. The detailed descriptions of the high-temperature and high-pressure experimental methods and measurement procedures have been reported in our previous work [35–37].

3. Results and Discussion

3.1. High-Pressure Raman Spectroscopy at Room Temperature

Figure 2 presents in situ Raman spectra of calcite with increasing pressures up to 19.4 GPa at atmospheric temperature. At the pressure of 0.5 GPa, the Raman spectrum of sample shows four Raman-active modes within the wavenumber range of 100–1200 cm^{-1}. According to previously reported Raman results on calcite [12], the strongest Raman characteristic peak at 1087.4 cm^{-1} is assigned to the symmetric stretching vibrational mode of carbonate ions (v_1). The weakest Raman peak of 714.4 cm^{-1} is related to the in-plane bend mode of carbonate ions (v_4) and two lattice modes of the translational (*T*) and librational (*L*) vibrations are observed at the respective positions of 155.4 cm^{-1} and 283.6 cm^{-1}. All those available Raman modes are in good agreement with previously reported Raman scattering results [9–11].

As seen from Figure 2a, within the pressure range of 0.5–1.1 GPa, all of the observed Raman peaks of sample tended to shift towards higher wavenumbers with increasing pressure. Upon compression to 1.6 GPa, the translational vibration mode (*T*) completely disappeared, which was accompanied by the emergence of three absolutely new Raman peaks at 129.5 cm^{-1} (M_1), 153.5 cm^{-1} (M_2), and 204.4 cm^{-1} (M_3). At the same time, the Raman peak of v_4 split into two Raman peaks at the respective positions of 709.7 cm^{-1} (M_4) and 724.5 cm^{-1} (v_4). These obvious Raman variations including the disappearance of the characteristic *T* mode, the appearance of three new Raman peaks (M_1, M_2, and M_3), and the splitting of the v_4 mode can provide robust evidences for the occurrence of structural transition from $CaCO_3$-I to $CaCO_3$-II phases, which is related to a strong rotational disorder of the CO_3 groups with clockwise rotation [38]. With the rise of pressure up to 2.2 GPa, the librational vibration mode (*L*) suddenly disappeared and seven new Raman peaks emerged at the respective positions of 108.7 cm^{-1} (M_5), 225.3 cm^{-1} (M_6), 275.1 cm^{-1} (M_7), 302.9 cm^{-1} (M_8), 345.1 cm^{-1} (M_9), 693.7 cm^{-1} (M_{10}), and 867.5 cm^{-1} (M_{11}). Simultaneously, a shoulder peak was observed at 1089.6 cm^{-1} (M_{12}) near the v_1 peak. In short, all of these clues from the variations of Raman characteristic peaks at 2.2 GPa reveal the occurrence of $CaCO_3$-II to $CaCO_3$-III structural phase transition by the displacement of Ca atom and the rotational disorder of CO_3 groups under high pressure [38]. In the pressure range of 2.2–14.2 GPa, the Raman peaks of sample continuously shifted towards higher wavenumbers except for the M_{11} Raman peak showing a red shift. As the pressure was increased to 16.8 GPa, the shoulder peak (M_{12}) and the v_4 peak suddenly disappeared, and a new Raman peak at 325.7 cm^{-1} (M_{13}) appeared in the lattice vibrational range of calcite. Meanwhile, the M_{11} mode exhibited a noticeable change from red shift to blue shift. These remarkable changes in Raman spectra indicate that $CaCO_3$-III phase transforms into $CaCO_3$-VI phase at 16.8 GPa. In addition, we also carefully checked the pressure-dependent Raman shifts of calcite during compression, as illustrated in Figure 3, and the corresponding pressure coefficients are clearly displayed in Table 1. It is noted that three obvious discontinuities in the pressure-dependent Raman shifts were detected at those of corresponding pressure points (i.e., 1.6, 2.2 and 16.8 GPa), which further denotes the pressure-induced structural phase transitions of calcite. In sum, our high-pressure Raman spectroscopic results reveal that calcite undergoes three structural phase transitions from $CaCO_3$-I to $CaCO_3$-II phases at 1.6 GPa, from $CaCO_3$-II to $CaCO_3$-III phases at 2.2 GPa, and from $CaCO_3$-III to $CaCO_3$-VI phases at 16.8 GPa upon compression, which are in excellent agreement with the reported Raman scattering results [9–11].

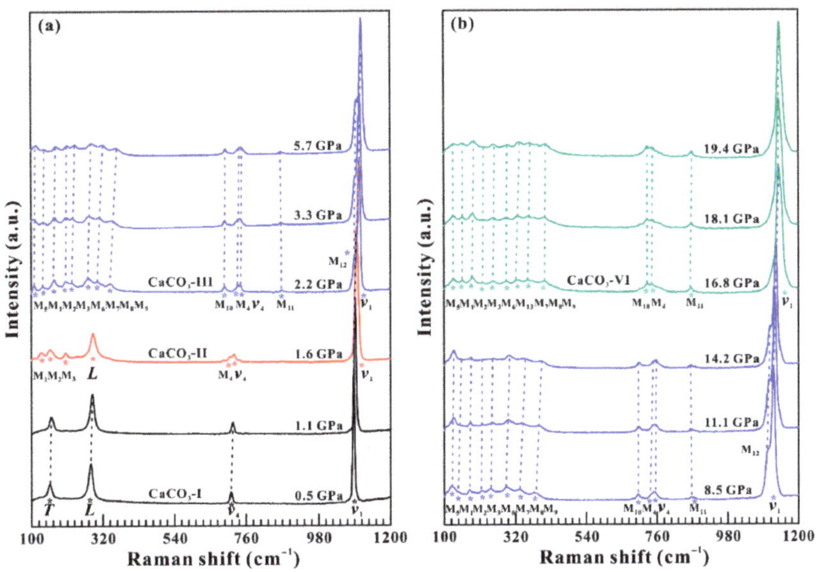

Figure 2. Raman scattering spectra of calcite at different pressure points up to 19.4 GPa upon compression at the atmospheric temperature. (**a**) 0.5–5.7 GPa; (**b**) 8.5–19.4 GPa. The red, blue, and green asterisks denote the positions of the Raman peaks of high-pressure $CaCO_3$ polymorphs. The colored dashed lines are used to guide the eyes.

Figure 4 displays in situ Raman spectra of calcite in the process of decompression from 18.3 GPa to 1 atm at room temperature. The corresponding pressure-dependent Raman shifts and pressure coefficients for calcite are illustrated in Figure 5 and Table 2, respectively. It is obvious that the Raman peaks of $CaCO_3$-VI phase exhibited red shift with decreasing pressure at 18.3–6.9 GPa. As the pressure was decompressed to 5.4 GPa, the characteristic Raman peaks of $CaCO_3$-III phase were recovered at the corresponding wavenumbers, implying the appearance of $CaCO_3$-VI to $CaCO_3$-III phase transition. Upon decompression to 1.5 GPa, the librational vibration mode (L) reemerged at 283.4 cm^{-1}, which reveals the structural phase transformation from $CaCO_3$-III to $CaCO_3$-II phases. Further releasing pressure to 0.4 GPa, the recoverable Raman peaks of $CaCO_3$-I phase are indicative of the onset of reverse structural transition from $CaCO_3$-II to $CaCO_3$-I phases. In brief, during decompression, the reverse structural phase transitions from $CaCO_3$-VI to $CaCO_3$-III to $CaCO_3$-II to $CaCO_3$-I phases are disclosed at the respective pressures of 5.4, 1.5, and 0.4 GPa. In comparison with the structural phase transitions of calcite at 1.6, 2.2, and 16.8 GPa in the compression process, a considerable pressure hysteresis of ~11 GPa between $CaCO_3$-VI and $CaCO_3$-III phases was detected, which is in good consistency with the obvious hysteresis of previously high-pressure infrared spectroscopy results on calcite to be reported by Catalli and Williams [39].

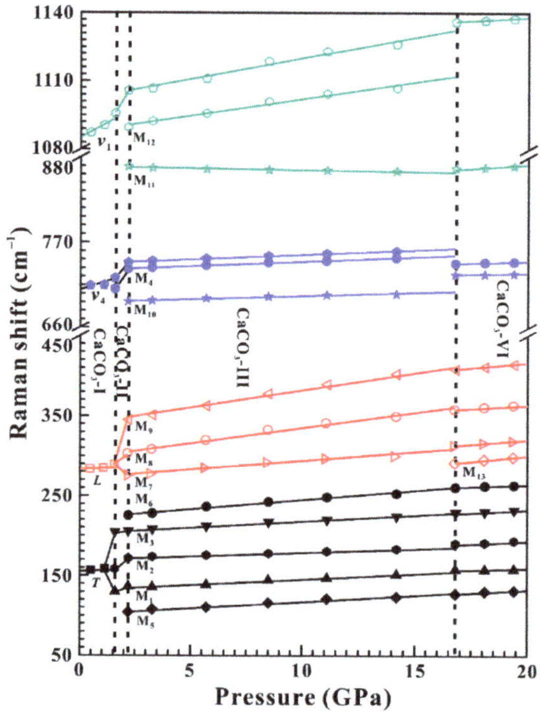

Figure 3. Pressure dependence of the Raman shifts of calcite upon compression within the wavenumber range of 100–1200 cm^{-1} at the atmospheric temperature. The colored solid lines and the vertical black dashed lines are used to guide the eyes.

3.2. *High-Pressure Electrical Conductivity at Ambient Temperature*

Previously available results have confirmed that the occurrence of pressure-induced phase transition for some representative minerals from the regions of the mid-lower Earth's crust and subduction zone also accompanied some discontinuous variations in electrical transport properties [28,29,31,40–43]. At room temperature, the in situ impedance spectroscopy experiments of calcite were measured up to 19.7 GPa to further shed light on its structural transitions, as plotted in Figure 6. It can be clearly seen from Figure 6a–c that the impedance spectra of calcite consist of a semicircular arc in the high frequency region (~10^3–10^7 Hz) and an oblique line in the low frequency range (10^{-1}–~10^3), which stands for the grain interior resistance (R_{gi}) and the grain boundary resistance (R_{gb}) for the calcite sample, respectively. To precisely determine the resistance values of the sample, ZView software was employed to fit the obtained impedance spectra. The fitting equivalent circuit is composed of two components in series and each component includes a resistance (R) and a constant phase element (CPE) in parallel. The fitting uncertainties of the impedance spectra were controlled to be less than 5%.

Table 1. Pressure coefficients of the Raman modes of $CaCO_3$-I, $CaCO_3$-II, $CaCO_3$-III, and $CaCO_3$-VI phases upon compression within the pressure range from 0.5–19.4 GPa at room temperature.

	Pressure Range (GPa)	Mode (cm^{-1})	Pressure Coefficient (cm^{-1} GPa^{-1})
$CaCO_3$-I phase	0.5–1.6 GPa	T	2.07
		L	1.83
		v_4	11.51
		v_1	16.14
$CaCO_3$-II phase	1.6–2.2 GPa	M_1	8.50
		M_2	22.83
		M_3	0.17
		M_4	42.03
		v_4	32.67
		v_1	16.50
$CaCO_3$-III phase	2.2–16.8 GPa	M_5	2.00
		M_1	1.38
		M_2	1.02
		M_3	1.61
		M_6	2.36
		M_7	2.28
		M_8	3.89
		M_9	5.09
		M_{10}	0.87
		M_4	1.19
		v_4	1.21
		M_{11}	–0.39
		M_{12}	1.46
		v_1	1.84
$CaCO_3$-VI phase	16.8–19.4 GPa	M_5	0.15
		M_1	0.50
		M_2	0.46
		M_3	0.76
		M_6	1.08
		M_{13}	2.81
		M_7	2.31
		M_8	1.62
		M_9	1.08
		M_{10}	0.65
		M_4	0.77
		M_{11}	0.88
		v_1	0.57

The electrical conductivity of calcite (σ) is calculated on the basis of the following equation:

$$\sigma = L/SR \tag{1}$$

where L is the experimental sample length (cm), S is the cross-sectional area of electrode (cm^2), and R is the resistance of sample (Ω), respectively. Figure 6d exhibits the pressure-dependent logarithmic electrical conductivity of calcite in both compressed and decompressed processes at room temperature. In light of the pressure-dependent electrical conductivity relation, three separate pressure regions were distinguished upon compression: (i) from 0.5 GPa to 2.3 GPa, the logarithmic electrical conductivity rapidly increased with a steep slope of 0.36 S cm^{-1} GPa^{-1}; (ii) from 2.3 GPa to 15.5 GPa, a gentle slope of 0.07 S cm^{-1} GPa^{-1} was acquired; (iii) from 15.5 GPa to 19.7 GPa, the logarithmic electrical conductivity values kept stable with a tiny slope of 0.03 S cm^{-1} GPa^{-1}. According to the high-pressure electrical conductivity results of calcite, two striking discontinuities at 2.3 GPa and 15.5 GPa were detected, which can be attributed to the pressure-induced structural phase transitions from $CaCO_3$-II to $CaCO_3$-III phases and $CaCO_3$-III to $CaCO_3$-

VI phases, respectively. Upon decompression, an apparent discontinuity in the electrical conductivity of calcite at 5.4 GPa can be ascribed to the reverse structural phase transition of $CaCO_3$-VI to $CaCO_3$-III phases. Furthermore, the recoverable electrical conductivity magnitude at 0.6 GPa indicates the reversibility of the structural transition of calcite.

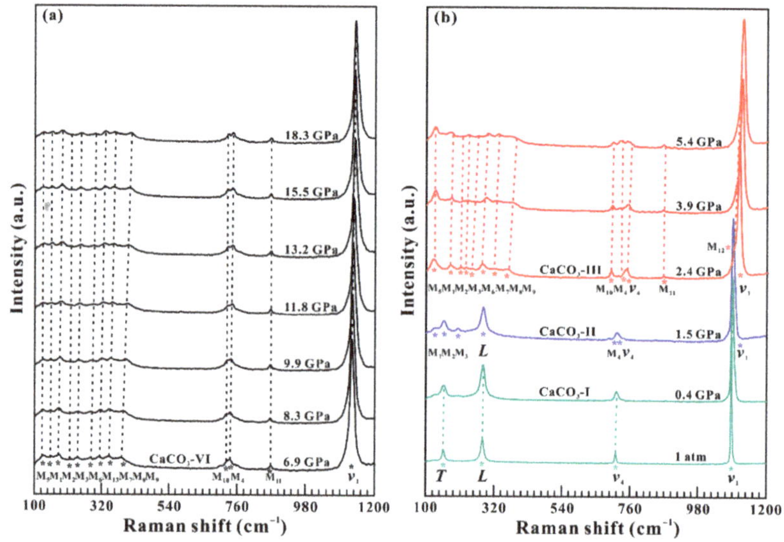

Figure 4. Raman scattering spectra of calcite upon decompression at the atmospheric temperature. (**a**) 18.3–6.9 GPa; (**b**) 5.4 GPa^{-1} atm. The colored dashed lines are used to guide the eyes.

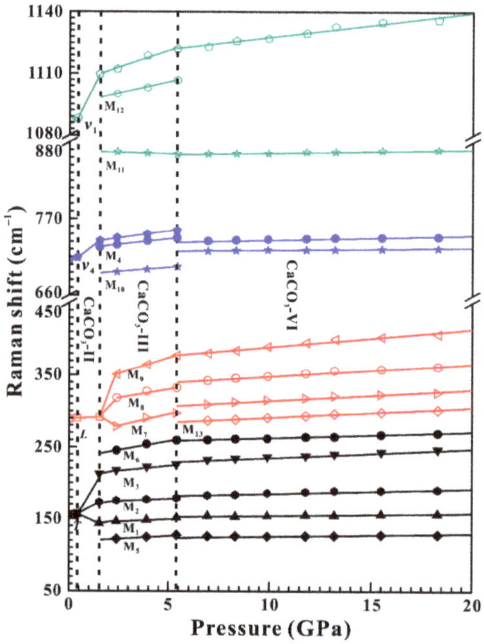

Figure 5. Pressure dependence of the Raman shifts of calcite upon decompression within the wavenumber range of 100–1200 cm^{-1} at the atmospheric temperature. The colored solid lines and the vertical black dashed lines are used to guide the eyes.

Table 2. Pressure coefficients of the Raman modes of $CaCO_3$-I, $CaCO_3$-II, $CaCO_3$-III, and $CaCO_3$-VI phases upon decompression within the pressure range from 18.3 GPa^{-1} atm at room temperature.

	Pressure Range (GPa)	Mode (cm^{-1})	Pressure Coefficient (cm^{-1} GPa^{-1})
$CaCO_3$-VI phase	18.3–5.4 GPa	M_5	0.20
		M_1	0.33
		M_2	0.73
		M_3	1.34
		M_6	0.87
		M_{13}	1.35
		M_7	1.54
		M_8	1.70
		M_9	2.47
		M_{10}	0.35
		M_4	0.50
		M_{11}	0.04
		v_1	1.18
$CaCO_3$-III phase	5.4–1.5 GPa	M_5	1.63
		M_1	1.33
		M_2	1.23
		M_3	2.87
		M_6	4.73
		M_7	6.37
		M_8	4.83
		M_9	8.87
		M_{10}	2.30
		M_4	3.27
		v_4	3.30
		M_{11}	−0.70
		M_{12}	2.23
		v_1	3.27
$CaCO_3$-II phase	1.5–0.4 GPa	M_1	7.48
		M_2	6.93
		M_3	8.27
		M_4	9.73
		v_4	10.26
		v_1	5.59
$CaCO_3$-I phase	0.4 GPa^{-1} atm	T	3.62
		L	1.21
		v_4	4.12
		v_1	3.72

In addition, Table 3 displays the structural phase transition pressures of calcite at room temperature obtained from previously reported results and ours for comparison. During compression, according to our presently obtained high-pressure Raman scattering and electrical conductivity results, calcite undergoes a series of structural transitions from $CaCO_3$-I to $CaCO_3$-II phases at 1.6 GPa, from $CaCO_3$-II to $CaCO_3$-III phases at 2.2 GPa, and from $CaCO_3$-III to $CaCO_3$-VI phases at 16.8 GPa, which accords well with the phase transition pressures of calcite in previous results reported by high-pressure Raman spectroscopy, Brillouin spectroscopy, synchrotron X-ray diffraction, shock wave experiment, and infrared spectroscopy measurements [9–12,38,39,44–46].

3.3. Electrical Conductivity at High Temperature and High Pressure

In order to explore the influence of temperature on the structural transition pressure between $CaCO_3$-III and $CaCO_3$-VI phases, a series of high-temperature electrical conductivity measurements were implemented at three representative pressure points of 10.5, 12.5, and 13.8 GPa. Figure 7a–c shows the impedance spectra of calcite in the temperature range of 323–873 K at a given pressure of 10.5 GPa. The corresponding logarithmic electrical conductivity as a function of reciprocal temperature is displayed in Figure 7d. The fitted parameters of the Arrhenius relation for the electrical conductivity of calcite at the pressure of 10.5, 12.5, and 13.8 GPa are displayed in a Table 4. Within the temperature ranges of 323–573 K and 723–873 K, the logarithmic electrical conductivity of the sample significantly increased with the rise of temperature, which is correlated with the enhanced mobility of charge carried at high temperatures [47]. However, a remarkable reduction in logarithmic electrical conductivity by approximately three orders of magnitude was observed at 573–723 K, which is probably related to the structural transition from $CaCO_3$-III to $CaCO_3$-VI phases. Specifically, at a fixed pressure of 10.5 GPa, the $CaCO_3$-III to $CaCO_3$-VI structural phase transition was completely ended at 723 K. Subsequently, we also measured the impedance spectra of calcite under conditions of temperature ranges of 323–873 K and 12.5 GPa, and the relationship between electrical conductivity and reciprocal temperature was established, as shown in Figure 8 in detail. Simultaneously, Figure 9 also displayed the impedance spectra of calcite, and its corresponding constructed relationship of logarithmic electrical conductivity with the variation of temperature within the fixed temperature range of 323–873 K and 13.8 GPa. Analogous to the above-mentioned Figure 7, all of these discontinuities in electrical conductivity with increasing temperature reveal the occurrences of the structural transformation from $CaCO_3$-III to $CaCO_3$-VI phases within the temperature ranges of 473–623 K and 423–523 K at the respective pressures of 12.5 GPa and 13.8 GPa. According to a series of detailed high-pressure electrical transport characterizations on calcite, we can initially outline that the $CaCO_3$-III to $CaCO_3$-VI structural phase transitions occurs at temperatures of 723, 623, and 523 K corresponding to three typical pressure points of 10.5, 12.5, and 13.8 GPa, respectively. In comprehensive combination with our previously obtained high-pressure Raman spectroscopy and electrical conductivity results at atmospheric temperature and high-temperature electrical conductivity results at high pressure, an obvious reducing tendency of phase transition temperature for calcite from the $CaCO_3$-III to $CaCO_3$-VI phases was observed with the rise of pressure.

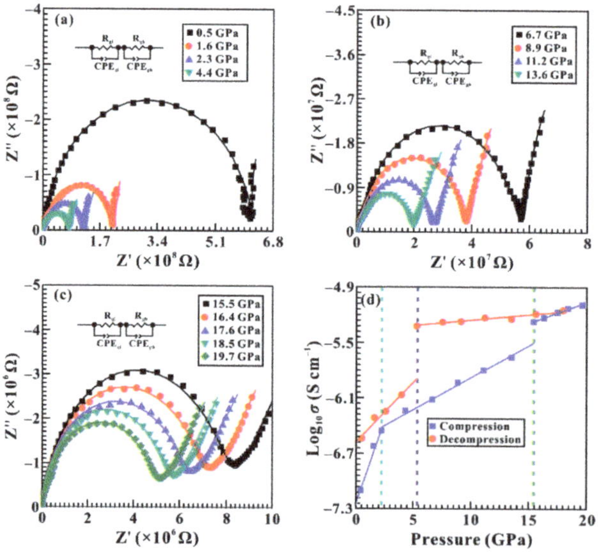

Figure 6. Complex impedance spectra of calcite at the atmospheric temperature within the pressure range of 0.5–19.7 GPa. (**a**) 0.5–4.4 GPa; (**b**) 6.7–13.6 GPa; (**c**) 15.5–19.7 GPa. (**d**) The pressure dependence of the logarithmic electrical conductivity of calcite in both processes of compression and decompression. Here, the symbols of Z′ and Z″ represent the real and imaginary parts of complex impedance spectra. The colored curves are the results of the fitting to the impedance spectra. The colored dashed lines and solid lines are employed to guide the eyes.

Table 3. Pressure-induced structural transition phases in calcite between $CaCO_3$-I and $CaCO_3$-II phases, $CaCO_3$-II and $CaCO_3$-III phases, and $CaCO_3$-III and $CaCO_3$-VI phases at atmospheric condition.

Phase Transition	Pressure (GPa)	Method	Reference
$CaCO_3$-I to $CaCO_3$-II phases	1.6	Raman spectroscopy	This study
	1.6	Raman spectroscopy	[9]
	1.6	Raman spectroscopy	[10]
	1.4	Raman spectroscopy	[11]
	1.7	Brillouin spectroscopy	[45]
	1.5	Synchrotron X-ray diffraction	[44]
$CaCO_3$-II to $CaCO_3$-III phases	2.3	Electrical conductivity	This study
	2.2	Raman spectroscopy	This study
	2.0	Raman spectroscopy	[9]
	2.0	Raman spectroscopy	[10]
	2.0	Raman spectroscopy	[12]
	2.1	Raman spectroscopy	[11]
	2.2	Brillouin spectroscopy	[45]
	2.5	Synchrotron X-ray diffraction	[44]
$CaCO_3$-III to $CaCO_3$-VI phases	15.5	Electrical conductivity	This study
	16.8	Raman spectroscopy	This study
	16.0	Raman spectroscopy	[12]
	15.5	Infrared spectroscopy	[39]
	15.0	Synchrotron X-ray diffraction	[44]

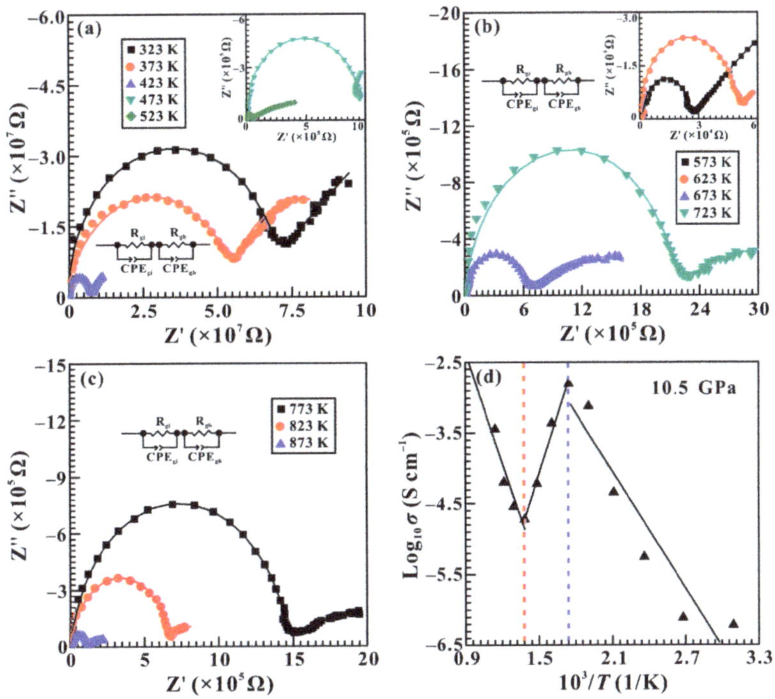

Figure 7. Complex impedance spectra of calcite in the temperature range of 323–873 K at a certain pressure point of 10.5 GPa. (**a**) 323–523 K; (**b**) 573–723 K; (**c**) 773–873 K. (**d**) The logarithmic electrical conductivity of calcite against reciprocal temperature. Z′ is the real part of complex impedance spectra and Z″ is the imaginary part. The colored curves are the results of the fitting to the impedance spectra. The black solid lines are obtained from the Arrhenius fitting results of the data.

Table 4. The fitted parameters for the Arrhenius relation of the electrical conductivity of calcite. *P* is the pressure, *T* is the absolute temperature, σ_0 is the pre-exponential factor, ΔH is the activation enthalpy, and R^2 is the linear correction.

P (GPa)	*T* (K)	Phase	σ_0 (S/cm)	ΔH (eV)	R^2
10.5	323–573	$CaCO_3$-III	5.89×10^1	5.48 ± 0.91	87.57
10.5	723–873	$CaCO_3$-VI	2.04×10^2	10.26 ± 2.71	82.08
12.5	323–473	$CaCO_3$-III	8.91×10^{-5}	1.62 ± 0.27	92.07
12.5	623–873	$CaCO_3$-VI	3.24×10^{-2}	5.90 ± 0.42	97.54
13.8	323–423	$CaCO_3$-III	1.73×10^{-3}	2.44 ± 1.01	70.46
13.8	523–873	$CaCO_3$-VI	1.77×10^0	6.55 ± 0.36	97.92

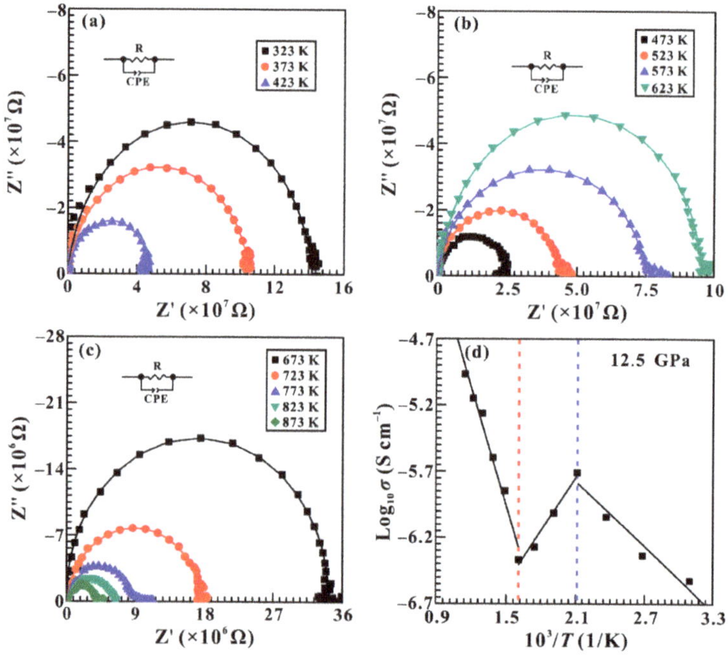

Figure 8. Complex impedance spectra of calcite in the temperature range of 323–873 K at 12.5 GPa. (**a**) 323–423 K; (**b**) 473–623 K; (**c**) 673–873 K. (**d**) The logarithmic electrical conductivity of calcite against reciprocal temperature. Z′ is the real part of the complex impedance spectra and Z″ is the imaginary part. The colored curves are the results of the fitting to the impedance spectra. The black solid lines are obtained from the Arrhenius fitting results of the data.

Figure 10 shows the pressure–temperature phase diagram between $CaCO_3$-III and $CaCO_3$-VI phases within the temperature range of 298–773 K over a wider pressure range from 1 atm to 16.0 GPa based on our acquired high-temperature and high-pressure electrical conductivity results. In here, the depth is fully considered according to the increasing pressure gradient value of 30 km/GPa in the deep Earth to be reported by Maeda et al. [6]. It is evident from the phase diagram of calcite that the phase transition pressures from $CaCO_3$-III to $CaCO_3$-VI phases decrease with the rise of temperature, and therefore, a negative dependence relation of phase transformation pressure on temperature was presented for the $CaCO_3$-III to $CaCO_3$-VI phase boundary. More importantly, the structural transition boundary between $CaCO_3$-III and $CaCO_3$-VI phases can be described as

$$P\ (\text{GPa}) = 19.219\ (\pm 1.105) - 0.011\ (\pm 0.002)\ T\ (\text{K}) \quad (2)$$

where *P* is the experimental pressure (GPa) and *T* is the experimental temperature (K). A negative phase transition slope between $CaCO_3$-III and $CaCO_3$-VI phases (–0.011 GPa K^{-1}) is observable, which is very close to the previously obtained value (–0.007 GPa K^{-1}) from the Raman spectroscopy measurements by Koch-Müller et al. [48]. In this study, the magnitude of higher absolute value in the slope indicates that the obtained phase transformation dependence relation of temperature on pressure is stronger by virtue of high-temperature and high-pressure electrical conductivity measurements. In addition, the Clausius-Clapeyron relation ($dT/dP = \Delta V/\Delta S$) has been widely applied to describe the correlation among entropy change (ΔS), molar volume change (ΔV) and phase transition slope (dT/dP) and to interpret the slope of structural transition of calcite under high-temperature and high-pressure conditions [46,49]. According to the previously available synchrotron X-ray results [44], $CaCO_3$-VI phase is denser than $CaCO_3$-III phase. Thus, the molar volume change (ΔV) would be a negative value before and after the phase transition. Entropy (*S*) is an important thermodynamic parameter of describing the disorder degree of a crystal. The high-pressure $CaCO_3$-VI phase with non-layered structure is more disordered than the $CaCO_3$-III phase containing planar CO_3 groups, leading to a positive value for the ΔS [44]. In comprehensive consideration of the negative ΔV and positive ΔS values, the phase transition boundary (dT/dP) between $CaCO_3$-III and $CaCO_3$-VI phases would be negative, which is consistent with our results obtained from high-temperature and high-pressure electrical conductivity measurements.

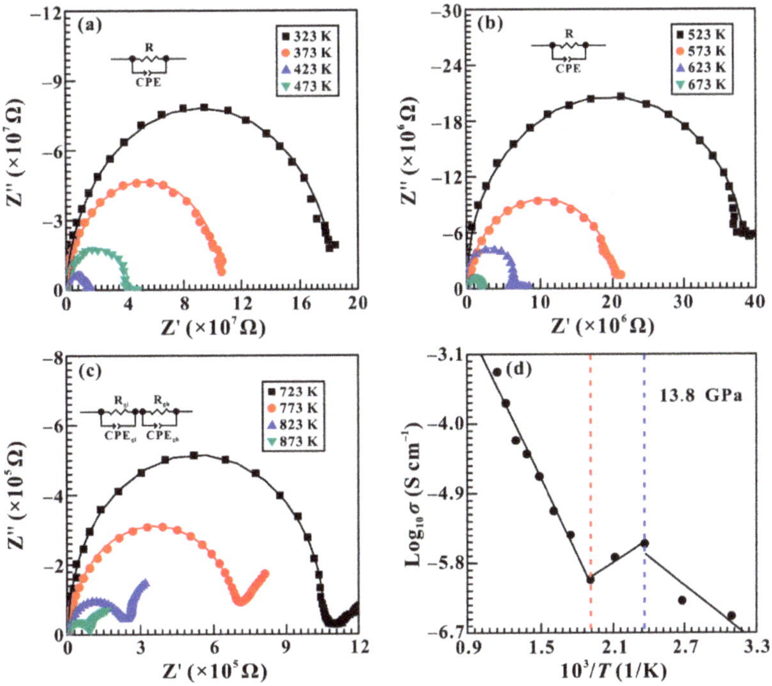

Figure 9. Complex impedance spectra of calcite in the temperature range of 323–873 K at 13.8 GPa. (**a**) 323–473 K; (**b**) 523–673 K; (**c**) 723–873 K. (**d**) The logarithmic electrical conductivity of calcite against reciprocal temperature. Z′ is the real part of the complex impedance spectra and Z″ is the imaginary part. The colored curves are the results of the fitting to the impedance spectra. The black solid lines are obtained from the Arrhenius fitting results of the data.

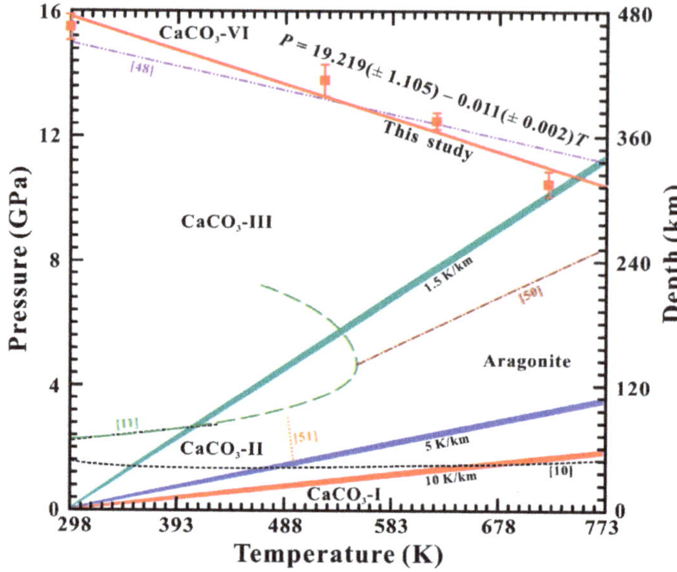

Figure 10. Temperature–pressure phase diagram of calcite according to the current and previous experimental investigations in the conditions of temperatures up to 773 K and pressures up to 16.0 GPa at the depth range of 0–480 km. The red solid squares are based on our present electrical conductivity experimental results. The red solid line is the fitted phase transition boundary between $CaCO_3$-III and $CaCO_3$-VI phases. The black dashed lines are the phase transition boundaries of $CaCO_3$-I to $CaCO_3$-II phases and $CaCO_3$-II to $CaCO_3$-III phases through high-temperature and high-pressure Raman scattering experimental results, coming from Yuan et al. [10]. The olive dashed curve comes from Suito et al. [11], standing for the phase transition boundary from $CaCO_3$-III phase to aragonite. The wine dot-dash line is the phase boundary from $CaCO_3$-III phase to aragonite by Bayarjargal et al. [50]. The orange dotted line represents the $CaCO_3$-II phase to aragonite boundary reported by Hess et al. [51]. The violet dot-dash line is the phase boundary between $CaCO_3$-III and $CaCO_3$-VI phases from Koch-Müller et al. [48]. The red, blue, and green areas represent the hot, cold, and very cold subducting slabs with the respective geothermal gradient values of 10 K/km, 5 K/km and 1.5 K/km.

4. Implications

In the present studies, we investigated the high-pressure phase stabilities of calcite, and a negative phase boundary between $CaCO_3$-III and $CaCO_3$-VI phases has been determined by means of Raman spectroscopy and electrical conductivity measurements under high-temperature and high-pressure conditions. Furthermore, a detailed phase diagram of calcite within the temperature range of 298–773 K and pressure of 0–16.0 GPa has been successfully established, and previously reported phase transition boundary results on calcite were also displayed in Figure 10 for comparison [9,10,48,50,51]. As seen in Figure 10, along the hot subducting slab with a geothermal gradient value of 10 K/km, calcite will transform into aragonite at the temperature of 650 K [10]. In the case of the cold subducting slab with a geothermal gradient value of 5 K/km, the $CaCO_3$-I phase first transforms into the $CaCO_3$-II phase at the temperature of 450 K, and further, the transformed $CaCO_3$-II phase turns into aragonite at the temperature of 500 K [10,51]. Significantly, along the very cold subducting slab with a geothermal gradient value of 1.5 K/km such as the Tonga subduction [52], the high-pressure structural transitions of calcite are more complex than those of the hot and cold subducting slabs. More specifically, along the very cold subducting slab, the structural phase transition from $CaCO_3$-I to $CaCO_3$-II phases appears at lower temperature of 350 K [10]. Subsequently, as the pressure increases up to 2.6 GPa, the $CaCO_3$-II phase is transformed into $CaCO_3$-III phase at the temperature of 400 K [10,11]. More importantly, our well-constructed phase boundary relationship between $CaCO_3$-III and $CaCO_3$-VI phases reveals that the transformed $CaCO_3$-III phase retains structural stability within the temperature range from 400 K to 750 K. Beyond the pressure of 10.8 GPa, the $CaCO_3$-VI phase emerges at the corresponding temperature of 750 K. Obviously, on the basis of previous results, it is possible that both of the $CaCO_3$-I and $CaCO_3$-II phases will transform into aragonite along the geothermal gradient values of 10 K/km and 5 K/km corresponding to the hot and cold slabs of subduction zone, respectively [10,51]. In conclusion, our new established phase diagram of the high-pressure polymorphs of $CaCO_3$-III and $CaCO_3$-VI phases at conditions of 298–773 K and 2.3–16.0 GPa has been well presented along the geothermal gradient value of 1.5 K/km corresponding to the very cold slab of subduction zone, which was similarly observed in the high-pressure $MgO–SiO_2–H_2O$ multi-phase equilibrium system along the geothermal gradient value of 1.5 K/km for the stagnant Tonga slab of a deeply subducted region up to 660 km [52].

Author Contributions: L.D. and H.H. supervised the project. L.D. and H.H. performed the conceptualization of this article and led the project. X.Z. and C.L. performed the investigation including X-ray diffraction (XRD), high-pressure Raman spectra, and high-pressure electrical conductivity measurements. L.D., H.H. and X.Z. contributed to the analysis, interpretation, and discussion of these results. X.Z., L.D. and H.H. performed the writing—original draft. L.D., H.H., X.Z. and C.L. performed the writing—review & editing. All authors have read and agreed to the published version of the manuscript.

Funding: This research was financially supported by the NSF of China (grant number 42072055 and 42274137) and the Youth Innovation Promotion Association of CAS (grant number 2019390).

Data Availability Statement: The data that support the findings of this study are available from the corresponding author upon reasonable request.

Acknowledgments: The authors acknowledge the technical support of the in situ high-pressure Raman scattering measurements provided by Heping Li at the Key Laboratory of High-temperature and High-pressure Study of the Earth's Interior, Institute of Geochemistry, Chinese Academy of Sciences.

Conflicts of Interest: The authors declare no conflict of interest.

References

1. Dasgupta, R.; Hirschmann, M.M. The deep carbon cycle and melting in the Earth's interior. *Earth Planet. Sci. Lett.* **2010**, *298*, 1–13. [CrossRef]
2. Dasgupta, R.; Hirschmann, M.M.; McDonough, W.F.; Spiegelman, M.; Withers, A.C. Trace element partitioning between garnet lherzolite and carbonatite at 6.6 and 8.6 GPa with applications to the geochemistry of the mantle and of mantle-derived melts. *Chem. Geol.* **2009**, *262*, 57–77. [CrossRef]
3. Doucet, L.S.; Li, Z.X.; El Dien, H.G. Oceanic and super-deep continental diamonds share a transition zone origin and mantle plume transportation. *Sci. Rep.* **2021**, *11*, 16958. [CrossRef] [PubMed]
4. Ghosh, S.; Ohtani, E.; Litasov, K.D.; Terasaki, H. Solidus of carbonated peridotite from 10 to 20 GPa and origin of magnesiocarbonatite melt in the Earth's deep mantle. *Chem. Geol.* **2009**, *262*, 17–28. [CrossRef]
5. Litasov, K.D.; Ohtani, E. Solidus and phase relations of carbonated peridotite in the system $CaO-Al_2O_3-MgO-SiO_2-Na_2O-CO_2$ to the lower mantle depths. *Phys. Earth Planet. Inter.* **2009**, *177*, 46–58. [CrossRef]
6. Maeda, F.; Ohtani, E.; Kamada, S.; Sakamaki, T.; Hirao, N.; Ohishi, Y. Diamond formation in the deep lower mantle: A high-pressure reaction of $MgCO_3$ and SiO_2. *Sci. Rep.* **2017**, *7*, 40602. [CrossRef] [PubMed]
7. Minarik, W.G.; Watson, E.B. Interconnectivity of carbonate melt at low melt fraction. *Earth Planet. Sci. Lett.* **1995**, *133*, 423–437. [CrossRef]
8. Tao, R.B.; Fei, Y.W. Recycled calcium carbonate is an efficient oxidation agent under deep upper mantle conditions. *Commun. Earth Environ.* **2021**, *2*, 45. [CrossRef]
9. Liu, L.G.; Mernagh, T.P. Phase transitions and Raman spectra of calcite at high pressures and room temperature. *Am. Mineral.* **1990**, *75*, 801–806.
10. Yuan, X.; Mayanovic, R.A.; Zhang, G. Phase transitions in $CaCO_3$ under hydrous and anhydrous conditions: Implications for the structural transformations of $CaCO_3$ during subduction processes. *Am. Mineral.* **2021**, *106*, 1780–1788. [CrossRef]

11. Suito, K.; Namba, J.; Horikawa, T.; Taniguchi, Y.; Sakural, N.; Kobayashi, M.; Onodera, A.; Shimomura, O.; Kikegawa, T. Phase relations of $CaCO_3$ at high pressure and high temperature. *Am. Mineral.* **2001**, *86*, 997–1002. [CrossRef]
12. Liu, J.; Caracas, R.; Fan, D.W.; Bobocioiu, E.; Zhang, D.Z.; Mao, W.L. High-pressure compressibility and vibrational properties of (Ca,Mn)CO_3. *Am. Mineral.* **2016**, *101*, 2723–2730. [CrossRef]
13. Bagdassarov, N.S.; Slutskii, A.B. Phase transformations in calcite from electrical impedance measurements. *Phase Transit.* **2003**, *76*, 1015–1028. [CrossRef]
14. Hu, H.Y.; Jing, C.X.; Dai, L.D.; Yin, C.Y.; Chen, D.M. Electrical conductivity of siderite and its implication for high conductivity anomaly in the slab-mantle wedge interface. *Front. Earth Sci.* **2022**, *10*, 985740. [CrossRef]
15. Mibe, K.; Ono, S. Electrical conductivity of $MgCO_3$ at high pressures and high temperatures. *Phys. B Condens. Matter* **2011**, *406*, 2018–2020. [CrossRef]
16. Ono, S.; Mibe, K. Electrical conductivity of aragonite in the subducted slab. *Eur. J. Mineral.* **2013**, *25*, 11–15. [CrossRef]
17. Ono, S.; Mibe, K. Influence of pressure and temperature on the electrical conductivity of dolomite. *Phys. Chem. Minerals.* **2015**, *42*, 773–779. [CrossRef]
18. Larson, A.C.; Von Dreele, R.B. General structure analysis system (GSAS). *Los Alamos Natl. Lab. Rep.* **2004**, *748*, 86–748.
19. Ishizawa, N.; Setoguchi, H.; Yanagisawa, K. Structural evolution of calcite at high temperatures: Phase V unveiled. *Sci. Rep.* **2013**, *3*, 2832. [CrossRef] [PubMed]
20. Maslen, M.; Streltsov, V.A.; Streltsova, N.R. X-ray study of the electron density in calcite, $CaCO_3$. *Acta Cryst.* **1993**, *B49*, 636–641. [CrossRef]
21. Mao, H.K.; Xu, J.; Bell, P.M. Calibration of the ruby pressure gauge to 800 kbar under quasi-hydrostatic conditions. *J. Geophys. Res. Solid Earth* **1986**, *91*, 4673–4676. [CrossRef]
22. Dai, L.D.; Wu, L.; Li, H.P.; Hu, H.Y.; Zhuang, Y.K.; Liu, K.X. Evidence of the pressure-induced conductivity switching of yttrium-doped $SrTiO_3$. *J. Phys. Condens. Matter* **2016**, *28*, 475501. [CrossRef]
23. Dai, L.D.; Wu, L.; Li, H.P.; Hu, H.Y.; Zhuang, Y.K.; Liu, K.X. Pressure-induced phase-transition and improvement of the microdielectric properties in yttrium-doped $SrZrO_3$. *Europhys. Lett.* **2016**, *114*, 56003. [CrossRef]
24. Hong, M.L.; Dai, L.D.; Hu, H.Y.; Zhang, X.Y.; Li, C. High-temperature and high-pressure phase transition of natural barite investigated by Raman spectroscopy and electrical conductivity. *Front. Earth Sci.* **2022**, *10*, 864183. [CrossRef]
25. Hong, M.L.; Dai, L.D.; Hu, H.Y.; Zhang, X.Y.; Li, C.; He, Y. Pressure-induced structural phase transition and metallization of $CrCl_3$ under different hydrostatic environments up to 50.0 GPa. *Inorg. Chem.* **2022**, *61*, 4852–4864. [CrossRef]
26. Hong, M.L.; Dai, L.D.; Hu, H.Y.; Zhang, X.Y.; Li, C.; He, Y. High-pressure structural phase transitions and metallization in layered HfS_2 under different hydrostatic environments up to 42.1 GPa. *J. Mater. Chem. C* **2022**, *10*, 10541–10550. [CrossRef]
27. Hu, H.Y.; Dai, L.D.; Sun, W.Q.; Zhuang, Y.K.; Liu, K.X.; Yang, L.F.; Pu, C.; Hong, M.L.; Wang, M.Q.; Hu, Z.M.; et al. Some remarks on the electrical conductivity of hydrous silicate minerals in the earth crust, upper mantle and subduction zone at high temperature and high pressures. *Minerals* **2022**, *12*, 161. [CrossRef]
28. Liu, K.X.; Dai, L.D.; Li, H.P.; Hu, H.Y.; Wu, L.; Zhuang, Y.K.; Pu, C.; Yang, L.F. Migration of impurity level reflected in the electrical conductivity variation for natural pyrite at high temperature and high pressure. *Phys. Chem. Miner.* **2018**, *45*, 85–92. [CrossRef]
29. Liu, K.X.; Dai, L.D.; Li, H.P.; Hu, H.Y.; Yang, L.F.; Pu, C.; Hong, M.L.; Liu, P.F. Phase transition and metallization of orpiment by Raman spectroscopy, electrical conductivity and theoretical calculation under high pressure. *Materials* **2019**, *12*, 784. [CrossRef]
30. Yang, L.F.; Dai, L.D.; Li, H.P.; Hu, H.Y.; Zhuang, Y.K.; Liu, K.X.; Pu, C.; Hong, M.L. Pressure-induced structural phase transition and dehydration for gypsum investigated by Raman spectroscopy and electrical conductivity. *Chem. Phys. Lett.* **2018**, *706*, 151–157. [CrossRef]
31. Yang, L.F.; Dai, L.D.; Li, H.P.; Hu, H.Y.; Hong, M.L.; Zhang, X.Y.; Liu, P.F. High-pressure investigations on the isostructural phase transition and metallization in realgar with diamond anvil cells. *Geosci. Front.* **2021**, *12*, 1031–1037. [CrossRef]
32. Zhang, X.Y.; Dai, L.D.; Hu, H.Y.; Hong, M.L. Pressure-induced metallic phase transition in gallium arsenide up to 24.3 GPa under hydrostatic conditions. *Mod. Phys. Lett. B* **2021**, *35*, 2150460. [CrossRef]
33. Zhang, X.Y.; Dai, L.D.; Hu, H.Y.; Hong, M.L.; Li, C. Pressure-induced coupled structural-electronic transition in SnS_2 under different hydrostatic environments up to 39.7 GPa. *RSC Adv.* **2022**, *12*, 2454–2461. [CrossRef]
34. Zhuang, Y.K.; Dai, L.D.; Wu, L.; Li, H.P.; Hu, H.Y.; Liu, K.X.; Yang, L.F.; Pu, C. Pressure-induced permanent metallization with reversible structural transition in molybdenum disulfide. *Appl. Phys. Lett.* **2017**, *110*, 122103. [CrossRef]
35. Dai, L.D.; Zhuang, Y.K.; Li, H.P.; Wu, L.; Hu, H.Y.; Liu, K.X.; Yang, L.F.; Pu, C. Pressure-induced irreversible amorphization and metallization with a structural phase transition in arsenic telluride. *J. Mater. Chem. C* **2017**, *5*, 12157–12162. [CrossRef]
36. Dai, L.D.; Liu, K.X.; Li, H.P.; Wu, L.; Hu, H.Y.; Zhuang, Y.K.; Yang, L.F.; Pu, C.; Liu, P.F. Pressure-induced irreversible metallization accompanying the phase transitions in Sb_2S_3. *Phys. Rev. B* **2018**, *97*, 024103. [CrossRef]
37. Dai, L.D.; Pu, C.; Li, H.P.; Hu, H.Y.; Liu, K.X.; Yang, L.F.; Hong, M.L. Characterization of metallization and amorphization for GaP under different hydrostatic environments in diamond anvil cell up to 40.0 GPa. *Rev. Sci. Instrum.* **2019**, *90*, 066103. [CrossRef] [PubMed]
38. Sivakumar, A.; Shailaja, P.; Nandhini, M.; Sahaya Jude Dhas, S.; Kumar, R.S.; Almansour, A.I.; Arumugam, N.; Chakraborty, S.; Martin Britto Dhas, S.A. Ternary switchable phase transition of $CaCO_3$ by shock waves. *Ceram. Int.* **2022**, *48*, 8457–8465. [CrossRef]
39. Catalli, K.; Williams, Q. A high-pressure phase transition of calcite-III. *Am. Mineral.* **2005**, *90*, 1679–1682. [CrossRef]

40. Hu, H.Y.; Dai, L.D.; Li, H.P.; Hui, K.S.; Sun, W.Q. Influence of dehydration on the electrical conductivity of epidote and implications for high-conductivity anomalies in subduction zones. *J. Geophys. Res. Solid Earth* **2017**, *122*, 2751–2762. [CrossRef]
41. Hu, H.Y.; Dai, L.D.; Li, H.P.; Sun, W.Q.; Li, B.S. Effect of dehydrogenation on the electrical conductivity of Fe-bearing amphibole: Implications for high conductivity anomalies in subduction zones and continental crust. *Earth Planet. Sci. Lett.* **2018**, *498*, 27–37. [CrossRef]
42. Hu, H.Y.; Dai, L.D.; Sun, W.Q.; Wang, M.Q.; Jing, C.X. Constraints on fluids in the continental crust from laboratory-based electrical conductivity measurements of plagioclase. *Gondwana Res.* **2022**, *107*, 1–12. [CrossRef]
43. Hong, M.L.; Dai, L.D.; Hu, H.Y.; Zhang, X.Y. Pressure-induced structural phase transitions in natural kaolinite investigated by Raman spectroscopy and electrical conductivity. *Am. Mineral.* **2022**, *107*, 385–394. [CrossRef]
44. Merlini, M.; Hanfland, M.; Crichton, W.A. $CaCO_3$-III and $CaCO_3$-VI, high-pressure polymorphs of calcite: Possible host structures for carbon in the Earth's mantle. *Earth Planet. Sci. Lett.* **2012**, *333*, 265–271. [CrossRef]
45. Zhao, C.S.; Li, H.P.; Chen, P.F.; Jiang, J.J. Sound velocities across calcite phase transition by Brillouin scattering spectroscopy. *Am. Mineral.* **2019**, *104*, 418–424. [CrossRef]
46. Zhao, X.; Zheng, Z.; Chen, J.Z.; Gao, Y.; Sun, J.H.; Hou, X.; Xiong, M.J.; Mei, S.H. High P-T calcite-aragonite phase transitions under hydrous and anhydrous conditions. *Front. Earth Sci.* **2022**, *10*, 907967. [CrossRef]
47. Trnovcová, V.; Furár, I.; Hanic, F. Influence of technological texture on electrical properties of industrial ceramics. *J. Phys. Chem. Solids* **2007**, *68*, 1135–1139. [CrossRef]
48. Koch-Müller, M.; Jahn, S.; Birkholz, N.; Ritter, E.; Schade, U. Phase transitions in the system $CaCO_3$ at high P and T determined by in situ vibrational spectroscopy in diamond anvil cells and first-principles simulations. *Phys. Chem. Miner.* **2016**, *43*, 545–561. [CrossRef]
49. Hou, M.Q.; Zhang, Q.; Tao, R.B.; Liu, H.; Kono, Y.; Mao, H.K.; Yang, W.G.; Chen, B.; Fei, Y.W. Temperature-induced amorphization in $CaCO_3$ at high pressure and implications for recycled $CaCO_3$ in subduction zones. *Nat. Commun.* **2019**, *10*, 1963. [CrossRef]
50. Bayarjargal, L.; Fruhner, C.J.; Schrodt, N.; Winkler, B. $CaCO_3$ phase diagram studied with Raman spectroscopy at pressures up to 50 GPa and high temperatures and DFT modeling. *Phys. Earth Planet. Inter.* **2018**, *281*, 31–45. [CrossRef]
51. Hess, N.J.; Ghose, S.; Exarhos, G.J. Raman spectroscopy at simultaneous high pressure and temperature: Phase relations of $CaCO_3$ and the lattice dynamics of the calcite $CaCO_3$(II) phase transition. In *Recent Trends in High Pressure Research; Proceedings of the X IIIth AIRAPT International Conference on High Pressure Science and Technology, Bangalore, India, 7–11 October 1991*; Singh, A.K., Ed.; US Department of Energy: Oak Ridge, TN, USA, 1991; pp. 236–241.
52. Komabayashi, T.; Omori, S.; Maruyama, S. Petrogenetic grid in the system MgO-SiO_2-H_2O up to 30 GPa, 1600 °C: Applications to hydrous peridotite subducting into the Earth's deep interior. *J. Geophys. Res. Solid Earth* **2004**, *109*, B03206. [CrossRef]

minerals

MDPI

Article

In Situ Raman Spectroscopy and DFT Studies of the Phase Transition from Zircon to Reidite at High P–T Conditions

Yue Gao [1,2], Zhi Zheng [1], Xia Zhao [3], Yuegao Liu [1], Jiangzhi Chen [1], Yan Li [4], Mengjun Xiong [1,2], Xiaotao Zu [5] and Shenghua Mei [1,*]

1 CAS Key Laboratory for Experimental Study under Deep-Sea Extreme Conditions, Institute of Deep-Sea Science and Engineering, Chinese Academy of Sciences, Sanya 572000, China
2 University of Chinese Academy of Sciences, Beijing 100049, China
3 College of Marine Science and Technology, Hainan Tropical Ocean University, Sanya 572022, China
4 Institute for Ocean Engineering, Shenzhen International Graduate School, Tsinghua University, Shenzhen 518055, China
5 School of Physics, University of Electronic Science and Technology of China, Chengdu 610054, China
* Correspondence: mei@idsse.ac.cn

Abstract: Zircon ($ZrSiO_4$) provides a good pressure-holding environment for ultra-high-pressure metamorphic minerals during crust exhumation due to its high incompressibility and chemical stability. At high pressure, the zircon can transform to reidite. Previous studies show much higher phase-transition pressures at room temperature than those at high temperature (>1000 K) due to kinetic hindrance. To further investigate the kinetics of the zircon–reidite phase transition at relatively low temperatures, the phase boundary at 298–800 K was determined using a diamond anvil cell combined with in situ Raman spectra. The results show that reidite becomes thermodynamically more stable compared with zircon at 8 GPa at room temperature, and the slope of the phase boundary at 298–800 K abruptly differs from that of previous studies at 1100–1900 K. Compared with the equilibrium phase boundary calculated by the density functional theory, it indicates that the kinetic effect of the zircon–reidite phase transition is obvious, and there exists a sufficiently large energy driving force provided by an overpressure to overcome the activation energy barrier below a critical temperature of approximately 880 K. The temperature dependence of overpressure is about 0.023 GPa/K.

Keywords: zircon; reidite; phase transition; high P–T experiment; diamond anvil cell; in situ Raman spectroscopy; density functional theory

Citation: Gao, Y.; Zheng, Z.; Zhao, X.; Liu, Y.; Chen, J.; Li, Y.; Xiong, M.; Zu, X.; Mei, S. In Situ Raman Spectroscopy and DFT Studies of the Phase Transition from Zircon to Reidite at High P–T Conditions. *Minerals* **2022**, *12*, 1618. https://doi.org/10.3390/min12121618

Academic Editors: Lidong Dai, Haiying Hu and Jordi Ibanez-Insa

Received: 24 October 2022
Accepted: 11 December 2022
Published: 15 December 2022

1. Introduction

Zircon is a ubiquitous mineral in many igneous, sedimentary, and metamorphic rocks [1]. It can accommodate radioactive trace elements and is therefore widely used in geochemical and geochronological studies [2–4]. Owing to its high incompressibility and chemical stability, zircon can form a good container for ultra-high-pressure metamorphic (UHPM) minerals during pressure release caused by the rapid exhumation of plates, leading to its application in exploring the geological pressure–temperature (P–T) record of the evolution of UHPM minerals [5–7]. Under ambient conditions, zircon exhibits tetragonal symmetry (space group $I4_1/amd$), whereas, under high pressure, zircon undergoes a first-order phase transition to reidite (scheelite structure, space group $I4_1/a$). Moreover, reidite has been found to be critical for understanding impact events [8–10]. For example, some reidite-bearing zircon samples were found in an upper Eocene impact ejecta layer in marine sediments on the upper continental slope off New Jersey and in Barbados [11–13]. Furthermore, zircon can be used as high-quality ceramics [14], refractories [15], and potential storage material for radioactive waste [16]. Therefore, understanding the phase transition of zircon to its high-pressure phase reidite and their physical properties under extreme conditions are of great significance in both geoscience and industries.

The zircon–reidite phase transition has been studied both experimentally [17–21] and theoretically [22–25]. Previous studies show that the phase transition pressure at room temperature is much higher than that at high temperature (>1000 K). For example, static compression experiments at room temperature indicated a transition pressure of 23 GPa via Raman spectroscopy [17] and 19.7 GPa via synchrotron radiation X-ray diffraction [20]. On the other hand, Ono et al. [18,19] reported that the zircon–reidite phase transition occurred at 8–10 GPa and 1100–1900 K by the multi-anvil apparatus and synchrotron radiation X-ray diffraction. In addition, Akaogi et al. [21] reported an equilibrium transition pressure of 8 GPa at 298 K by thermodynamic calculations. Based on the first principle calculations, the zircon–reidite equilibrium boundary is located around 5.2–7.3 GPa at 0 K [23–25]. The much higher transition pressure compared to the equilibrium transition pressure under room temperature suggests that a substantial activation barrier exists for the zircon–reidite transition. Interestingly, the zircon–reidite phase transition at room temperature exhibits "an anomalous and possibly unique silicate transformation mechanism" [17], since first-order phase transitions at room temperature are unusual among silicates. Regarding phase transition mechanics, Kusaba et al. [26,27] proposed a two-step mechanism, and Marques et al. [23] proposed a quasi-monoclinic transition path. The phase transition activation energy decreases with increasing pressure; thus, the transition can occur at room temperature.

However, the kinetics of the zircon–reidite phase transition is not well constrained experimentally at relatively low temperatures (298–1000 K). To refine the high P–T phase diagram of zircon–reidite, we determined the kinetic phase boundary between zircon and reidite in the range of 300–800 K using a diamond anvil cell (DAC) in conjunction with in situ Raman spectroscopy. Furthermore, combined with the thermodynamic equilibrium phase boundary calculated by the density functional theory (DFT), the influence of kinetic hindrance on the zircon–reidite phase transition was investigated quantitatively.

2. Materials and Methods

We selected natural zircons from a meta-gabbro sample from the Langmuri platinum group elements deposit in the Eastern Kunlun Orogenic Belt. Zircon grains were separated from bulk samples using conventional heavy liquid and magnetic techniques and then purified by handpicking under a binocular microscope.

High-pressure experiments were performed using a BX90 DAC [28] with a culet size of 500 μm in diameter. Zircon samples were manually ground and polished to a size of ~60 × 60 × 20 μm^3. Then, they were loaded into a chamber (200 μm in diameter) that was drilled by laser into a pre-indented rhenium gasket to about 40–50 μm thickness. A 16:3:1 mixture of methanol/ethanol/water was used as the pressure transmitting medium to ensure a quasi-hydrostatic pressure environment. The fluorescence of 5 at.% doped Sm:YAG was used to monitor the pressure of the sample chamber by using the pressure-dependent shift of its Y1 fluorescence line [29]:

$$P(\text{GPa}) = (\text{A}/\text{B}) \times \{[1 + (\Delta\lambda/\lambda_0)]^{\text{B}} - 1\} \quad (1)$$

where A = 2089.91(23.04), B = −4.43(1.07), $\Delta\lambda = \lambda - \lambda_0$, and λ and λ_0 are the wavelengths of the Y1 line at ambient pressure and elevated pressure in nanometers, respectively. The sample chamber was heated with an external resistive (Ni-Cr) heater, and the temperature was measured using a K-type thermocouple (CHAL-010, Omega Engineering Inc., Stamford, CT, USA) attached to the diamond anvils.

For the convenience of pressure loading under high temperatures, the DAC was integrated into a piezoelectric ceramic pressure loading system. In the system, when an input electric signal is applied to the piezoelectric ceramic using a function generator and a power amplifier, the piezoelectric ceramic expands and compresses the DAC in a controlled and continuous manner. In this study, a stepped voltage-time signal was used to drive the piezoelectric ceramic with 3 stages, rising, constant, and falling, corresponding to compression, pressure-holding, and decompression, respectively. In the experiments, the sample was first slowly heated (25 K/min) to the target temperature; then, the input signal was applied to the piezoelectric ceramic, and the pressure loading procedure commenced. Since the low signal-to-noise ratio of Raman signal at high P–T, the experiments were carried out in situ below 500 K and ex situ above 500 K, respectively. In the in situ Raman spectra of the sample and the fluorescence spectra of Sm:YAG were simultaneously obtained 12 min after each pressure-holding stage began, and in the ex situ Raman experiments, the samples were held at the chosen pressure for 20 min. After decompression, the temperature was reduced to room temperature at a rate of 25 K/min, and the Raman spectra of the samples for the ex situ experiments were observed at room temperature and atmospheric pressure.

The confocal micro-Raman spectroscopy equipment comprises two spectrometers (Andor Shamrock 500i) equipped with a CCD detector (Andor iVac 316) and a 532 nm laser with 251 mW laser power (MLL-FN-532/1–800 mW, Changchun New Industries Optoelectronics Technology Co. Ltd., Changchun, China) [30]. A spectrometer was used to record the Raman signal of zircon with a 10 s exposure time, a 100 μm aperture slit, and an 1800 g/mm grating affording a resolution of 3.02 cm^{-1}. The other spectrometer was used to record the fluorescence signal of Sm:YAG with 0.1 s exposure time, 10 μm aperture slit, and 1200 g/mm grating affording a resolution of 0.06 nm. Before each experiment, two spectrometers were thoroughly calibrated using a standard mercury lamp.

The thermodynamic equilibrium phase boundary of zircon–reidite was calculated by DFT. To account for the temperature and pressure dependence of the Gibbs free energy, the quasi-harmonic approximation method [31,32] was employed to calculate the total Helmholtz free energy $F(T;V)$, including the internal energy $U(V)$ and the phonon Helmholtz free energy $F_{phonon}(T;V)$ of zircon and reidite at 11 cell volume points (V) with temperatures (T) ranging from 0 to 1900 K in steps of 2 K. DFT calculations were performed using the Vienna ab initio simulation package (VASP) [33,34]. The exchange–correlation interactions were treated with the local density approximation (LDA) functional [35]. The valence–core interactions were described using the projector–augmented wave method [36] with a cutoff energy of 600 eV for the plane–wave basis set, where the valence electron configurations of Zr (4s4p5s4d), Si (s2p2), and O (s2p4) were selected. The convergence criteria were 10^{-6} eV and -10^{-4} eV/Å for electronic and ionic relaxations, respectively. The PHONOPY code [32] was applied to obtain the phonon properties using the density functional perturbation theory methods [37] with a 9 × 9 × 9 k-point mesh for the primitive unit cell. The Vinet equation of state [38] was used to fit the F–V points to the curves, the minimums of which correspond to the equilibrium volumes and the Helmholtz free energy. Additionally, the Gibbs free energy $G(T,p)$ at a given temperature T and the pressure p was obtained from the Helmholtz free energy $F(T;V)$ through the following transformation:

$$G(T,p) = \min_V [F(T;V) + \mathrm{pV}] \quad (2)$$

3. Results and Discussion

We conducted a series of high pressure experiments at different temperatures (300–800 K) to build the zircon–reidite phase boundary. All experimental conditions and results are summarized in Table 1. As shown in Figure 1, the Raman spectrum of the initial zircon sample in the experiment DD0159 contains seven peaks at 202, 215, 225, 356, 440, 975, and 1006 cm^{-1}, which represent the $E_g(1)$, $B_{1g}(1)$, $E_g(2)$, $E_g(3)$, $A_{1g}(1)$, $A_{1g}(2)$, and $B_{1g}(4)$ vibrational modes of zircon, respectively. This feature is in good agreement with previous experimental and computational results [39–42]. The other spectra in Figure 1 (DD0115, DD0162, and DD0159) represent the recovered samples after experiments under different P–T conditions. The typical Raman vibrational modes of reidite observed in our experiments and related assignments are listed in Table 2, which well correspond to the results of former researchers [40,41,43]. Furthermore, the Raman peaks of the recovered sample in DD0115 and DD0162 shift by about 3–10 cm^{-1} toward high wave numbers, in contrast to those in DD0159, and the reasons have not yet been identified.

Table 1. Experimental conditions and results.

Run	Temperature (K)	Pressure (GPa)	Zircon–Reidite Transition Pressure (GPa)
(In situ Raman observation at high temperature and high pressure)			
DD0079	298	1.2–7.6	— [1]
DD0080 [2]	298	6.6–15.1	—
DD0091	298	9.2–19.5	19.1
DD0114	298	4.4–18.4	—
DD0115	298	11.4–26.0	19.5
DD0123	298	13.6–25.0	21.0
DD0214	298	0.6–21.8	20.0
DD0162(1)	373	6.5–14.3	—
DD0162(2) [3]	373	10.4–25.0	18.7
DD0159	473	13.7–25.0	16.0
Run	**Temperature (K)**	**Pressure (GPa)**	**Recovered Samples**
(Ex situ Raman observation at room temperature and atmospheric pressure)			
CD0155	298	24.0	zircon + reidite
CD0031	373	8.7	zircon
CD0039	373	13.2	zircon
CD0032	473	9.4	zircon
CD0154	500	16.1	zircon + reidite
CD0033	573	8.7	zircon
CD0172(1) [4]	600	12.5	zircon
CD0172(2)	700	10.0	zircon
CD0172(3)	800	15.2	zircon + reidite

[1] No phase transition was observed; [2] DD0079 and DD0080 are two cycles of one pressure-loading procedure using the same sample; [3] DD0162 (1) and (2) are two cycles of one pressure-loading procedure using the same sample; [4] CD0172 (1), (2), and (3) represent three experiments performed sequentially using the same sample.

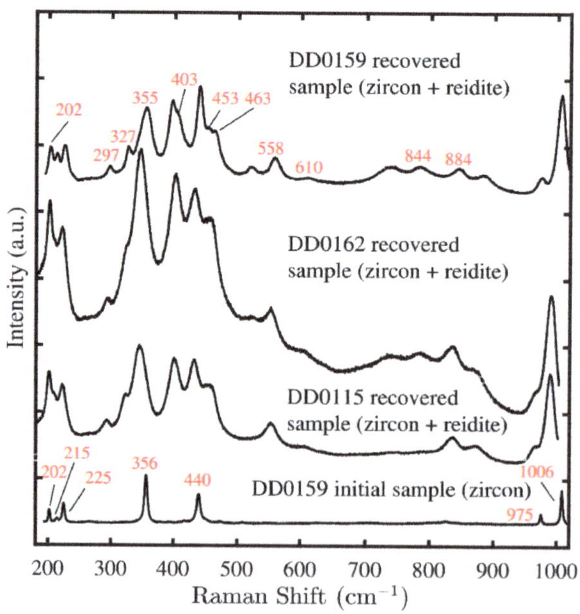

Figure 1. Raman spectra of the initial and recovered samples in the experiments. Due to the overlap of the peak, we cannot distinguish the peak at 202 cm^{-1} as belonging to reidite or zircon in the spectrum of recovered sample DD0159.

Table 2. Raman shifts of reidite.

Vibrational Modes	ν_0 [1] (cm^{-1})	ν_0 [2] (cm^{-1})	ν_0 [3] (cm^{-1})	ν_0 [4] (cm^{-1})	ν_0 [5] (cm^{-1})
E_g (1)	202	196 (+6) [6]	204 (−2)	—	209 (−7)
B_g (1)	—	—	238	237.9	242
E_g (2)	297	—	297 (0)	297.0 (0)	300 (−3)
A_g (1)	327	320 (+7)	327 (0)	326.4 (−0.6)	326 (+1)
B_g (2)	355	343 (+12)	353 (+2)	349 (+6)	350 (+5)
A_g (2)	403	—	406 (−3)	406.1 (−3.1)	409 (−6)
E_g (3)	453	—	—	—	458 (−5)
B_g (3)	463	456 (+7)	464 (−1)	465.8 (−2.8)	465 (−2)
E_g (4)	558	552 (+6)	558 (0)	558.4 (−0.4)	558 (0)
B_g (4)	610	604	610 (0)	610.9 (−0.9)	608 (+2)
B_g (5)	844	842 (+2)	847 (−3)	847 (−3)	852 (−8)
A_g (3)	—	—	—	—	861
E_g (5)	884	880 (+4)	887 (−3)	886.4 (−2.4)	891 (−7)

[1] Data were from DD0159. [2] Data were from Knittle and Williams [17]. [3] Data were derived from Gucsik [43]. [4] Data were reported by Mihailova et al. [40]. [5] Data source was Stangarone et al. [41]. [6] The numbers in brackets represent the difference in the vibration frequency from the values in DD0159.

The characteristics of the zircon Raman spectra at different pressures are listed in Figure 2. In DD0115, new peaks were observed at 508 and 611 cm^{-1} at 19.5 GPa; they are labeled by blue and red arrows in Figure 2a and are attributed to the B_g(3) and E_g(4) vibrational modes of reidite, respectively, while these two vibrational modes are located at 463 and 558 cm^{-1} under ambient conditions. Likewise, in DD0162(2), the B_g(3) and E_g(4) modes of the reidite were observed at 18.7 GPa, as denoted with blue and red arrows in Figure 2b, respectively. In DD0159, the E_g(4) modes of reidite was observed at 16.0 GPa, as denoted with a red arrow in Figure 2c, and the B_g(3) mode was not observed, possibly due to the low signal-to-noise ratio at high P–T. Our results indicate that zircon–reidite phase transitions occur at 19.5, 18.7, and 16.0 GPa when the temperature is 298, 373, and 473 K, respectively.

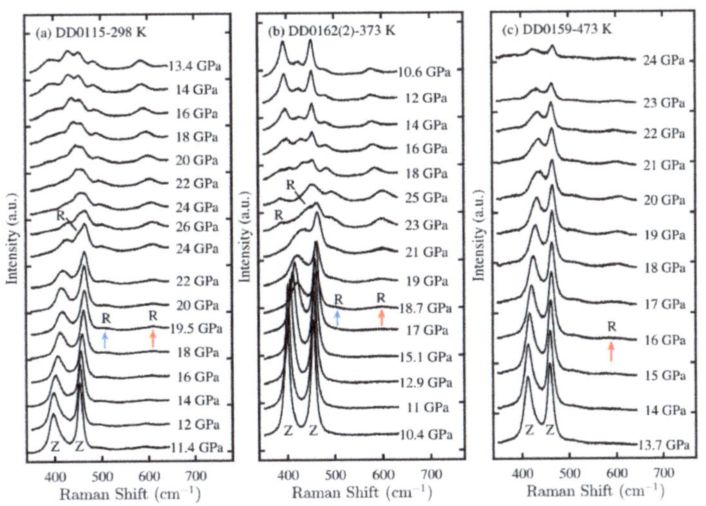

Figure 2. The characteristics of zircon Raman spectra at different pressures for (**a**) DD0115, (**b**) DD0162(2), and (**c**) DD0159. The peaks of zircon and reidite are marked by Z and R, respectively. The blue and red arrows denote the novel peaks that are attributed to the vibrational modes of reidite B_g(3) and E_g(4), respectively.

Before the zircon–reidite phase transition occurs, the Raman spectra of zircon show some anomalous features with increasing pressure. Figure 3a,b display the Raman spectra of the sample in DD0079 and DD0080, corresponding to the pressure ranges of 1.2–7.6 and 6.6–15.1 GPa, respectively, at room temperature. Figure 3c summarizes the pressure dependence of the wavenumber of the $B_{1g}(1)$ (215 cm^{-1} under ambient conditions), $E_g(1)$ (202 cm^{-1} under ambient conditions), and $E_g(2)$ (224 cm^{-1} under ambient conditions) modes. As shown by the red arrows in Figure 3a,b, the $B_{1g}(1)$ and $E_g(2)$ modes merge into a peak at 6.1 GPa. As the pressure further increases, it splits into two peaks at 8.8 GPa. These phenomena can be attributed to the crossing of the $B_{1g}(1)$ and $E_g(2)$ modes due to the larger pressure dependence of the wavenumber $d\omega/dP$ of the $B_{1g}(1)$ mode than that of the $E_g(2)$ mode. In addition, in Figure 3a, the intensity of the $E_g(2)$ mode relative to the $B_{1g}(1)$ mode decreases with increasing pressure, resulting in the intensity of the $B_{1g}(1)$ mode being greater than that of the $E_g(2)$ mode in Figure 3b. At about 8 GPa, the value of $d\omega/dP$ of the $E_g(2)$ mode changes from slightly positive to negative, while the value of $d\omega/dP$ of the $B_{1g}(1)$ mode remains relatively unchanged, as shown in Figure 3c. Furthermore, the split of the $E_g(1)$ mode into two peaks is indicated with a blue arrow in Figure 3b at 10.6 GPa, which is caused by the sudden decrease of $d\omega/dP$ of the $E_g(1)$ mode of some zircons at around 8 GPa, as shown in Figure 3c, while the other zircons maintain its original $d\omega/dP$ value. The abrupt changes of $d\omega/dP$ of the $E_g(1)$ and $E_g(2)$ modes are related to the change of the interatomic force constants [40], resulting in the thermodynamic metastability of zircon above 8 GPa below room temperature, which is generally consistent with the previously experimental results (10 GPa) by in situ Raman [40].

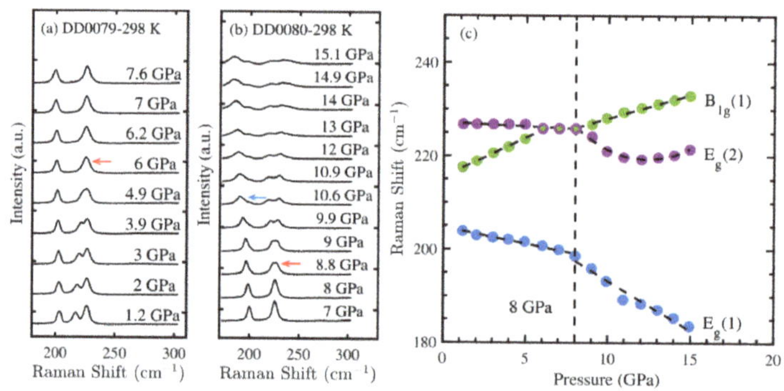

Figure 3. The characteristics of zircon Raman spectra at different pressures for (**a**) DD0079 and (**b**) DD0080. The red arrow in (**a**) represents the merger of the $B_{1g}(1)$ and $E_g(2)$ modes into one peak, which splits into two peaks, as shown by the red arrow in (**b**). The blue arrow in (**b**) represents the split of the $E_g(1)$ mode into two peaks. (**c**) The Raman shift of the $E_g(1)$, $B_{1g}(1)$, and $E_g(2)$ zircon modes as a function of pressure for DD0079 and DD0080. The vertical dashed lines correspond to the abrupt changes of $d\omega/dP$ of the $E_g(1)$ and $E_g(2)$ modes.

Additionally, a similar variation of the $B_{1g}(1)$ and the $E_g(2)$ modes in the zircon–reidite phase transition at high temperatures was observed (Figure 4). The results indicate that zircon becomes thermodynamically metastable at about 8 GPa/373 K and 11.1 GPa/473 K.

Moreover, we calculated the thermodynamic equilibrium phase boundary of zircon–reidite in the 0–1900 K temperature range using DFT. By comparing the Gibbs free energy at a fixed temperature and different pressures, the thermodynamic stability differences between the two phases were obtained. When $\Delta G = 0$, the pressure at this point corresponds to the equilibrium phase transition pressure under the current temperature. For example, the phase transition pressures were 5.69, 6.53, 7.65, and 8.70 GPa at temperatures of 0, 500, 1000, and 1500 K, respectively (Figure 5).

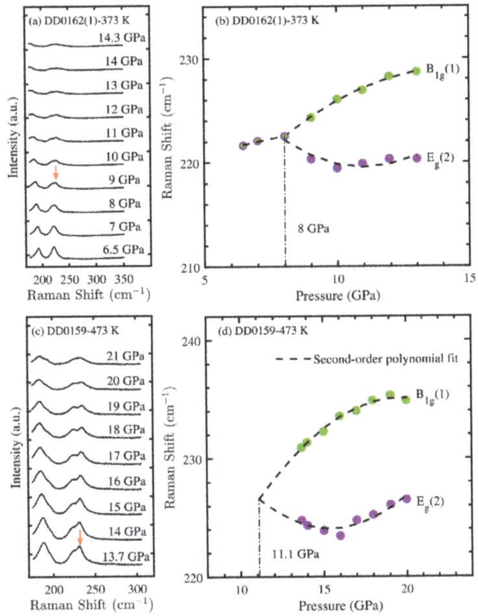

Figure 4. The characteristics of zircon Raman spectra at different pressures in (**a**) DD0162(1) and (**c**) DD0159. The red arrows represent the split of the peak containing the $B_{1g}(1)$ and $E_g(2)$ modes. Raman shift of the zircon $B_{1g}(1)$ and $E_g(2)$ modes at different pressures in (**b**) DD0162(1) and (**d**) DD0159. The dashed lines represent the second-order polynomial fit to the data. The vertical dot-dashed lines correspond to the abrupt change of $d\omega/dP$ of the $E_g(2)$ modes.

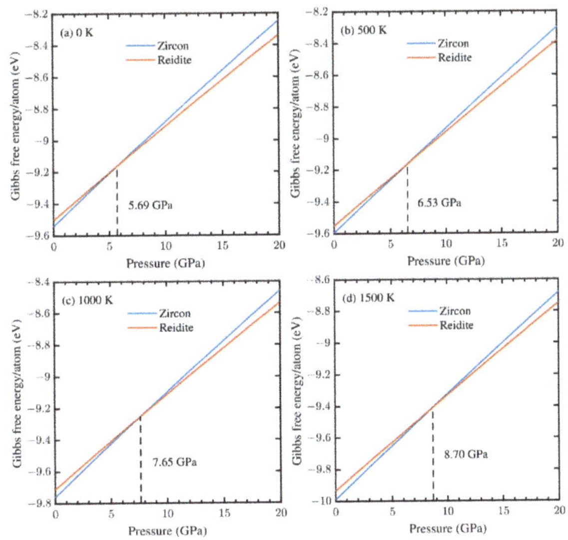

Figure 5. The Gibbs free energy of zircon and reidite unit cells as a function of pressure at 0 K (**a**), 500 K (**b**), 1000 K (**c**) and 1500 K (**d**). The vertical dashed lines correspond to the phase transition pressures.

Figure 6 compares the high P–T phase boundary of zircon–reidite determined through experiments and calculations in this study with those of previous studies. Our experimental results show the phase transition pressure of zircon–reidite at room temperature is 19.5 GPa, which is consistent with that (19.7 GPa) reported by van Westrenen et al. [20] via DAC and synchrotron radiation X-ray diffraction but 3 GPa lower than that (23 GPa) reported by Knittle and Williams [17] via DAC and Raman. The reason for their higher transition pressure might be the fact that their experimental samples are the metamorphic zircon, which contains more impurities than the magmatic zircon we used. Our calculated equilibrium phase boundary by DFT using the LDA functional agrees well with the equilibrium phase boundary by the multi-anvil apparatus and synchrotron radiation X-ray diffraction between 1100 and 1900 K from Ono et al. [18] and has essentially the same slope as the results by thermodynamic calculations between 298 and 1900 K from Akaogi et al. [21]. Our in situ Raman experimental results suggest reidite becomes thermodynamically more stable compared with zircon at 8 GPa, which is close to the calculational equilibrium transition pressure (about 6 GPa) by us and the experimental results (about 10 GPa) by Mihailova et al. [40] at room temperature. Other DFT calculations show similar equilibrium transition pressures around 5.2, 5.3 and 7.3 GPa at 0 K reported by Dutta and Mandal [25], Marques et al. [23], and Du et al. [24], respectively.

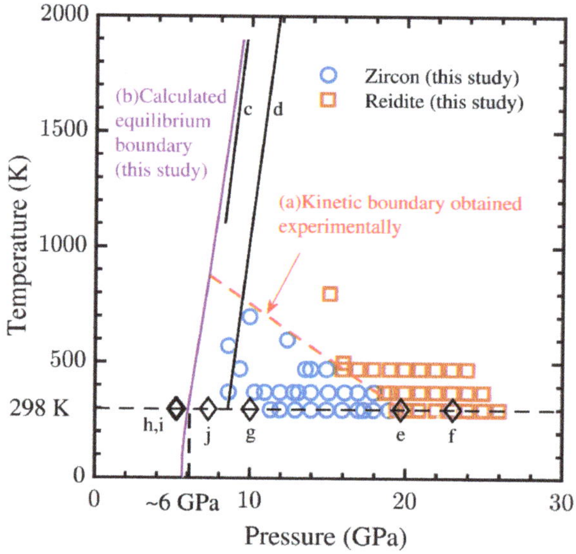

Figure 6. Experimental and computational high P–T results for the zircon–reidite phase transition. The blue circles and orange squares represent the stable domains of zircon and reidite, respectively. Abbreviations are as follows: (a) the kinetic phase boundary determined by linear fit of our experiments results; (b) the equilibrium phase boundary calculated at 298 to 1900 K by DFT in this study; (c) the equilibrium phase boundary at 1100 to 1900 K experimentally determined by Ono et al. [18]; (d) the equilibrium phase boundary at 298 to 2000 K from Akaogi et al. [21]; (e,f) the phase transition pressure at room temperature experimentally determined by van Westrenen et al. [20] and Knittle and Williams [17], respectively; (g) the pressure above which zircon becomes thermodynamically metastable with respect to the reidite reported by Mihailova et al. [40]; and (h–j) the calculated equilibrium transition pressures from Dutta and Mandal [25], Marques et al. [23], and Du et al. [24], respectively.

According to the numerical fitting of experimental points, the zircon–reidite phase boundary is approximately linear, with a negative Clapeyron slope in the temperature range of 298–800 K, as indicated by the red dashed line in Figure 6. The experimentally obtained phase boundary will intersect with the calculated equilibrium phase boundary (shown as the purple solid line in Figure 6) at a critical temperature of approximately 850 K, which divides the phase transition into two stages: (1) above 880 K, the phase transition occurs under thermodynamic equilibrium conditions where the phase boundary has a positive

dP/dT slope, and (2) below 880 K, the slope of dP/dT changes from positive to negative. Compared with the calculated equilibrium phase boundary, it indicates that there exists an overpressure (ΔP) at low temperatures to provide a sufficiently large energy driving force (ΔG_f) and to overcome the kinetic energy barrier that is caused by small displacements of atoms within a two-step model for the zircon–reidite phase transition proposed by Kusaba et al. [27]. Thus, the zircon–reidite phase boundary obtained experimentally at 298–800 K represents a kinetic boundary. Moreover, the driving force for the zircon–reidite phase transition was calculated using the following formula [44]:

$$\Delta G_f = \Delta V \Delta P \quad (3)$$

where ΔV is the cell volume change upon the phase transition obtained by DFT calculations. Considering the various values of ΔV at different pressures, Formula (3) should be modified as follows:

$$\Delta G_f = \int_{P_e}^{P_k} \Delta V \mathrm{d}P \quad (4)$$

where P_e and P_k are equilibrium transition pressures and kinetic transition pressures, respectively. In addition, the activation energy (ΔG) affected by the overpressure can be overcome by an available vibrational energy, which can be calculated by subtracting the zero point energy from the Gibbs free energy at finite temperatures from DFT calculational results.

As listed in Table 3, the values of ΔG_f and ΔG were calculated at each temperature. It is obvious that ΔG_f increases with the decreasing temperature, resulting in a decrease of ΔG. The temperature dependence of overpressure is about 0.023 GPa/K, which is consistent with the value of ~0.02 GPa/K estimated from Figure 1 of [23]. When the temperature is higher than about 880 K, ΔG_f is close to 0, and the vibrational energy is sufficiently large to overcome the activation energy barrier. Since the barrier height of the zircon–reidite phase transition is mainly affected by pressure, with negligible influence of temperature [23], the sum of ΔG_f and ΔG will be equal to a constant. However, in Table 3, we can note that the values of $\Delta G_f + \Delta G$ are in the range of 58.4–68.9 kJ/mol below 573 K, while those above 673 K are relatively larger. Therefore, the actual kinetic phase boundary may be located under the red dashed line in Figure 6 above 673 K, and the critical temperature may be relatively lower than 880 K.

Table 3. The values of ΔG_f and ΔG.

Temperature T (K)	Kinetic Transition Pressure P_k (GPa)	Equilibrium Transition Pressure P_e (GPa)	Driving Force ΔG_f (kJ/mol)	Activation Energy ΔG (kJ/mol)
298	19.5 [1]	6.1	49.0	9.4
373	18.7	6.3	45.3	16.2
473	16.0	6.5	34.9	27.9
573	14.0 [2]	6.7	26.8	42.1
673	11.8	6.9	18.0	58.7
773	9.7	7.1	9.6	77.4
873	7.5	7.3	0.7	98.0

[1] The values of P_k at 298, 373, and 473 K were obtained from our experiments. [2] The values of P_k at the temperature above 473 K were obtained by the linear fit of those at 298, 373, and 473 K.

4. Conclusions

In summary, our research reveals that the zircon–reidite phase transition is significantly hindered by kinetic factors. The slope of the experimentally determined kinetic phase boundary differs significantly from that of the calculated equilibrium phase boundary at 298–800 K. There exists an overpressure to provide the sufficiently large energy driving force and to overcome the kinetic energy barrier below a critical temperature of approximately 880 K, and temperature dependence of overpressure is about 0.023 GPa/K.

Author Contributions: Conceptualization, Y.G., Z.Z., Y.L. (Yuegao Liu), J.C. and S.M.; Data curation, Y.G. and J.C.; Formal analysis, Y.G. and Z.Z.; Funding acquisition, S.M.; Investigation, Y.G., X.Z. (Xia Zhao) and Y.L. (Yan Li); Methodology, Y.G., Z.Z., X.Z. (Xia Zhao), Y.L. (Yan Li) and M.X.; Project administration, S.M.; Resources, Y.G., Z.Z., X.Z. (Xia Zhao), Y.L. (Yan Li) and M.X.; Software, Y.G., J.C. and X.Z. (Xiaotao Zu); Supervision, Z.Z., Y.L. (Yuegao Liu), J.C. and S.M.; Validation, Z.Z., X.Z. (Xia Zhao), Y.L. (Yuegao Liu) and J.C.; Visualization, Y.G. and J.C.; and Writing—original draft, Y.G.; Writing—review and editing, Y.G., Z.Z., Y.L. (Yuegao Liu), J.C. and S.M. All authors have read and agreed to the published version of the manuscript.

Funding: This research was funded by the Chinese Academy of Sciences, grant numbers QYZDY-SSW-DQC029 and XDA22040501.

Data Availability Statement: Not applicable.

Acknowledgments: We are grateful to the anonymous reviewers for their helpful comments.

Conflicts of Interest: The authors declare no conflict of interest.

References

1. Finch, R.J.; Hanchar, J.M.; Hoskin, P.W.O.; Burns, P.C. Rare-earth elements in synthetic zircon: Part 2. A single-crystal X-ray study of xenotime substitution. *Am. Miner.* **2001**, *86*, 681–689. [CrossRef]
2. Hawkesworth, C.; Kemp, A. Using hafnium and oxygen isotopes in zircons to unravel the record of crustal evolution. *Chem. Geol.* **2006**, *226*, 144–162. [CrossRef]
3. Lissenberg, C.J.; Rioux, M.; Shimizu, N.; Bowring, S.A.; Mével, C. Zircon Dating of Oceanic Crustal Accretion. *Science* **2009**, *323*, 1048–1050. [CrossRef] [PubMed]
4. Gao, Y.J.; Liu, Y.G.; Zhang, J.W.; Kong, H.L.; Chen, A.P.; Kou, S.L. Post-UHP tectonic evolution of the East Kunlun Orogenic Belt, northern Tibetan Plateau: Insight from the regional-scale crustal anatexis. *Int. Geol. Rev.* **2022**, 2059707. [CrossRef]
5. Rubatto, D.; Hermann, J. Zircon Behaviour in Deeply Subducted Rocks. *Elements* **2007**, *3*, 31–35. [CrossRef]
6. Korhonen, F.J.; Clark, C.; Brown, M.; Taylor, R.J.M. Taking the temperature of Earth's hottest crust. *Earth Planet. Sci. Lett.* **2014**, *408*, 341–354. [CrossRef]
7. Liu, P.; Jin, Z. Metamorphic Evolution of a Tremolite Marble from the Dabie UHP Terrane, China: A Focus on Zircon. *J. Earth Sci.* **2022**, *33*, 493–506. [CrossRef]
8. Wittmann, A.; Kenkmann, T.; Schmitt, R.T.; Stöffler, D. Shock-metamorphosed zircon in terrestrial impact craters. *Meteorit. Planet. Sci.* **2006**, *41*, 433–454. [CrossRef]
9. Cavosie, A.J.; Erickson, T.M.; Timms, N.E. Nanoscale records of ancient shock deformation: Reidite ($ZrSiO_4$) in sandstone at the Ordovician Rock Elm impact crater. *Geology* **2015**, *43*, 315–318. [CrossRef]
10. Tomioka, N.; Miyahara, M. High-pressure minerals in shocked meteorites. *Meteorit. Planet. Sci.* **2017**, *52*, 2017–2039. [CrossRef]
11. Thein, J. A Tektite Layer in Upper Eocene Sediments of the New Jersey Continental Slope (Site 612, Leg 95). *Initial. Rep. Deep. Sea Drill. Proj.* **1987**, *95*, 565–579. [CrossRef]
12. Glass, B.P.; Liu, S.B. Discovery of high-pressure $ZrSiO_4$ polymorph in naturally occurring shock-metamorphosed zircons. *Geology* **2001**, *29*, 371–373. [CrossRef]
13. Glass, B.P.; Liu, S.; Leavens, P.B. Reidite: An impact-produced high-pressure polymorph of zircon found in marine sediments. *Am. Miner.* **2002**, *87*, 562–565. [CrossRef]
14. Scott, H.P.; Williams, Q.; Knittle, E. Ultralow Compressibility Silicate without Highly Coordinated Silicon. *Phys. Rev. Lett.* **2002**, *88*, 015506. [CrossRef]
15. Gucsik, A.; Zhang, M.; Koeberl, C.; Salje, E.K.H.; Redfern, S.A.T.; Pruneda, J.M. Infrared and Raman spectra of $ZrSiO_4$ experimentally shocked at high pressures. *Miner. Mag.* **2004**, *68*, 801–811. [CrossRef]
16. Ewing, R.C. Nuclear waste forms for actinides. *Proc. Natl. Acad. Sci. USA* **1999**, *96*, 3432–3439. [CrossRef]
17. Knittle, E.; Williams, Q. High-pressure Raman spectroscopy of $ZrSiO_4$: Observation of the zircon to scheelite transition at 300 K. *Am. Mineral.* **1993**, *78*, 245–252.
18. Ono, S.; Funakoshi, K.; Nakajima, Y.; Tange, Y.; Katsura, T. Phase transition of zircon at high P-T conditions. *Contrib. Miner. Pet.* **2004**, *147*, 505–509. [CrossRef]
19. Ono, S.; Tange, Y.; Katayama, I.; Kikegawa, T. Equations of state of $ZrSiO_4$ phases in the upper mantle. *Am. Miner.* **2004**, *89*, 185–188. [CrossRef]
20. Van Westrenen, W.; Frank, M.R.; Hanchar, J.M.; Fei, Y.W.; Finch, R.J.; Zha, C.-S. In situ determination of the compressibility of synthetic pure zircon ($ZrSiO_4$) and the onset of the zircon-reidite phase transition. *Am. Miner.* **2004**, *89*, 197–203. [CrossRef]
21. Akaogi, M.; Hashimoto, S.; Kojitani, H. Thermodynamic properties of $ZrSiO_4$ zircon and reidite and of cotunnite-type ZrO2 with application to high-pressure high-temperature phase relations in $ZrSiO_4$. *Phys. Earth Planet. Inter.* **2018**, *281*, 1–7. [CrossRef]
22. Marqués, M.; Flórez, M.; Recio, J.M.; Gerward, L.; Olsen, J.S. Structure and stability of $ZrSiO_4$ under hydrostatic pressure. *Phys. Rev. B* **2006**, *74*, 014104. [CrossRef]

23. Marqués, M.; Contreras-García, J.; Flórez, M.; Recio, J.M. On the mechanism of the zircon-reidite pressure induced transformation. *J. Phys. Chem. Solids* **2008**, *69*, 2277–2280. [CrossRef]
24. Du, J.C.; Devanathan, R.; Corrales, L.R.; Weber, W.J. First-principles calculations of the electronic structure, phase transition and properties of $ZrSiO_4$ polymorphs. *Comput. Theor. Chem.* **2012**, *987*, 62–70. [CrossRef]
25. Dutta, R.; Mandal, N. Structure, elasticity and stability of reidite ($ZrSiO_4$) under hydrostatic pressure: A density functional study. *Mater. Chem. Phys.* **2012**, *135*, 322–329. [CrossRef]
26. Kusaba, K.; Syono, Y.; Kikuchi, M.; Fukuoka, K. Shock behavior of zircon: Phase transition to scheelite structure and decomposition. *Earth Planet. Sci. Lett.* **1985**, *72*, 433–439. [CrossRef]
27. Kusaba, K.; Yagi, T.; Kikuchi, M.; Syono, Y. Structural considerations on the mechanism of the shock-induced zircon-scheelite transition in $ZrSiO_4$. *J. Phys. Chem. Solids* **1986**, *47*, 675–679. [CrossRef]
28. Kantor, I.; Prakapenka, V.; Kantor, A.; Dera, P.; Kurnosov, A.; Sinogeikin, S.; Dubrovinskaia, N.; Dubrovinsky, L. BX90: A new diamond anvil cell design for X-ray diffraction and optical measurements. *Rev. Sci. Instrum.* **2012**, *83*, 125102. [CrossRef]
29. Trots, D.M.; Kurnosov, A.; Ballaran, T.B.; Tkachev, S.; Zhuravlev, K.; Prakapenka, V.; Berkowski, M.; Frost, D.J. The Sm:YAG primary fluorescence pressure scale. *J. Geophys. Res. Solid Earth* **2013**, *118*, 5805–5813. [CrossRef]
30. Zhao, X.; Zheng, Z.; Chen, J.Z.; Gao, Y.; Sun, J.H.; Hou, X.; Xiong, M.J.; Mei, S.H. High P-T Calcite-Aragonite Phase Transitions Under Hydrous and Anhydrous Conditions. *Front. Earth Sci.* **2022**, *10*, 907967. [CrossRef]
31. Togo, A.; Chaput, L.; Tanaka, I.; Hug, G. First-principles phonon calculations of thermal expansion in Ti3SiC2, Ti3AlC2, and Ti3GeC2. *Phys. Rev. B* **2010**, *81*, 174301. [CrossRef]
32. Togo, A.; Tanaka, I. First principles phonon calculations in materials science. *Scr. Mater.* **2015**, *108*, 1–5. [CrossRef]
33. Kresse, G.; Furthmüller, J. Efficient iterative schemes forab initiototal-energy calculations using a plane-wave basis set. *Phys. Rev. B* **1996**, *54*, 11169–11186. [CrossRef]
34. Kresse, G.; Joubert, D. From ultrasoft pseudopotentials to the projector augmented-wave method. *Phys. Rev. B* **1999**, *59*, 1758–1775. [CrossRef]
35. Perdew, J.P.; Zunger, A. Self-interaction correction to density-functional approximations for many-electron systems. *Phys. Rev. B* **1981**, *23*, 5048–5079. [CrossRef]
36. Blöchl, P.E. Projector augmented-wave method. *Phys. Rev. B Condens. Matter Mater. Phys.* **1994**, *50*, 17953–17979. [CrossRef]
37. Baroni, S.; de Gironcoli, S.; Dal Corso, A.; Giannozzi, P. Phonons and related crystal properties from density-functional perturbation theory. *Rev. Mod. Phys.* **2001**, *73*, 515–562. [CrossRef]
38. Vinet, P.; Rose, J.H.; Ferrante, J.; Smith, J.R. Universal features of the equation of state of solids. *J. Phys. Condens. Matter* **1989**, *1*, 1941–1963. [CrossRef]
39. Binvignat, F.A.P.; Malcherek, T.; Angel, R.J.; Paulmann, C.; Schlüter, J.; Mihailova, B. Radiation-damaged zircon under high pressures. *Phys. Chem. Miner.* **2018**, *45*, 981–993. [CrossRef]
40. Mihailova, B.; Waeselmann, N.; Stangarone, C.; Angel, R.J.; Prencipe, M.; Alvaro, M. The pressure-induced phase transition(s) of $ZrSiO_4$: Revised. *Phys. Chem. Miner.* **2019**, *46*, 807–814. [CrossRef]
41. Stangarone, C.; Angel, R.J.; Prencipe, M.; Mihailova, B.; Alvaro, M. New insights into the zircon-reidite phase transition. *Am. Miner.* **2019**, *104*, 830–837. [CrossRef]
42. Smirnov, M.B.; Mirgorodsky, A.P.; Kazimirov, V.Y.; Guinebretière, R. Bond-switching mechanism for the zircon-scheelite phase transition. *Phys. Rev. B* **2008**, *78*, 094109. [CrossRef]
43. Gucsik, A. Micro-Raman spectroscopy of reidite as an impact-induced high pressure polymorph of zircon: Experimental investigation and attempt to application. *Acta Mineral. Petrogr.* **2007**, *47*, 17–24.
44. Zhang, J.; Li, B.; Utsumi, W.; Liebermann, R.C. In situ X-ray observations of the coesite-stishovite transition: Reversed phase boundary and kinetics. *Phys. Chem. Miner.* **1996**, *23*, 1–10. [CrossRef]

minerals

MDPI

Article

Sound Velocity Measurement of Shock-Compressed Quartz at Extreme Conditions

Liang Sun [1], Huan Zhang [1], Zanyang Guan [1], Weiming Yang [1], Youjun Zhang [2], Toshimori Sekine [3,*], Xiaoxi Duan [1,*], Zhebin Wang [1] and Jiamin Yang [1]

1 Laser Fusion Research Center, Chinese Academy of Engineering Physics, Mianyang 621900, China; sun.is.liang@gmail.com (L.S.); helenzhang1986@gmail.com (H.Z.); gzy0707@mail.ustc.edu.cn (Z.G.); ywm835093137@yeah.net (W.Y.); zhebinw@vip.sina.com (Z.W.); yjm70018@sina.com (J.Y.)

2 Institute of Atomic and Molecular Physics, Sichuan University, Chengdu 610065, China; zhangyoujun@scu.edu.cn

3 Center for High Pressure Science and Technology Advanced Research (HPSTAR), Shanghai 201203, China

* Correspondence: toshimori.sekine@hpstar.ac.cn (T.S.); xiaoxi_duan@163.com (X.D.)

Abstract: The physical properties of basic minerals such as magnesium silicates, oxides, and silica at extreme conditions, up to 1000 s of GPa, are crucial to understand the behaviors of magma oceans and melting in Super-Earths discovered to data. Their sound velocity at the conditions relevant to the Super-Earth's mantle is a key parameter for melting process in determining the physical and chemical evolution of planetary interiors. In this article, we used laser indirectly driven shock compression for quartz to document the sound velocity of quartz at pressures of 270 GPa to 870 GPa during lateral unloadings in a high-power laser facility in China. These measurements demonstrate and improve the technique proposed by Li et al. [PRL 120, 215703 (2018)] to determine the sound velocity. The results compare favorably to the SESAME EoS table and previous data. The Grüneisen parameter at extreme conditions was also calculated from sound velocity data. The data presented in our experiment also provide new information on sound velocity to support the dissociation and metallization for liquid quartz at extreme conditions.

Keywords: laser shock compression; sound velocity; high-pressure; quartz; Grüneisen parameter; super-earth

Citation: Sun, L.; Zhang, H.; Guan, Z.; Yang, W.; Zhang, Y.; Sekine, T.; Duan, X.; Wang, Z.; Yang, J. Sound Velocity Measurement of Shock-Compressed Quartz at Extreme Conditions. *Minerals* **2021**, *11*, 1334. https://doi.org/10.3390/min11121334

Academic Editors: Julien Siebert, Lidong Dai, Haiying Hu and Jianjun Jiang

Received: 30 October 2021
Accepted: 16 November 2021
Published: 28 November 2021

1. Introduction

Thousands of exoplanets such as CoRoT or Kepler outside our Solar system were discovered, raising fundamental questions about unique planetary formation mechanism [1–3] and their corresponding interior structures and dynamics [4]. Among these numerous planets, a significant proportion are to a large degree made of silica and silicate, although not all are entirely rocky [5–7]. Among these constituents, silica (SiO_2) play an important role in stable magma oceans, which are relevant to dynamo and magnetic field generation and heat transfer as well as the habitability for exoplanets [8,9]. Characterizing the equation of state (EoS), phase diagram, metallization, and dissociation process of silica at extreme conditions in exoplanets is of evident important for understanding the properties of magma oceans [10] and melting in Super-Earth interior [11].

The high-pressure behavior of silica was the subject of extensive experimental studies [12–16]. A number of stable crystalline structure and metastable forms of SiO_2 were observed experimentally or predicted theoretically [17,18], but pressure is limited in most studies to below 200 GPa [19,20]. The properties of silica in Super-Earth interiors are still poorly constrained regarding the extreme conditions in pressure and temperature.

Laboratory laser driven shock loading presents a unique capability to yield TPa pressure-temperature states comparable to planetary interior [1,21]. As an impedance-matching standard material, the principle Hugoniot [22–25] and release isentrope [26,27]

of α−quartz were studied extensively in the fluid region. Shock-induced vaporization of quartz occurs at much lower pressures than previous estimates because of the entropy on the Hugoniot [28]. Melting temperatures up to 8300 K above 500 GPa for fused silica, α-quartz and stishovite were reported in laser-driven shock experiments [29]. For further increases in pressure along the Hugoniot, silica transitions form a bonded molecular liquid to a dissociated atomic fluid provided by specific heat c_v information [24], and the evolution of the electronic structure [30]. The metallization and dissociation in silica influence strongly the equation of state and evolution of the associated transport properties.

Sound velocity describes the off-Hugoniot properties of a material and offers constraints on presence of chemical bonds based on a thermodynamically consistent analysis. The experimental determination of the speed of sound behind a shock front is of great interest for studying phase transition and geophysical problems [31–34]. The Grüneisen coefficient could be also determined by the sound velocity along a known Hugoniot curve. The study of sound velocity for silica in high pressure by dynamic shock experiments could provide further information on dissociation and metallization and increase our confidence for our theoretical models for Super-Earth interior condition.

In this article, we present high-pressure sound velocity and Grüneisen coefficient by laser shock compression experiment for α-quartz. These experiments revisited and provided absolute measurement of sound velocity in α−quartz from 200 to 900 GPa by improving the previous technique proposed by Li et al. [35]. The experimental results are compared to SESAME EoS table, in which there are various theoretical and experimental results by different platforms for various silica polymorphs.

2. Methods

2.1. Sound Velocity Method by Edge Rarefaction

Methods for measuring the speed of sound behind a shock front were developed by Al'tshuler et al. [36,37]. Method of lateral unloading is one of them, in which disturbances from the unloading overtake the shock front and weaken the shock. The front velocity in the weakened outer section of the front surface decreases and the shock velocity image on the streak curves, while the central part not yet disturbed remains planar. The trace of the bent location provides the speed of sound and the shock velocity as a function of transit time. The sound velocity c_s can be obtained by

$$c_s = u_s \times (\tan^2 \alpha + (\frac{u_s - u_p}{u_s})^2)^{1/2}, \tag{1}$$

where u_s is shock velocity and α is unloading angle between the shock direction and the tangent of bending trance. u_p is the particle velocity behind shock. But the location where shock front bends and sound velocity is difficult to determine by experiments.

The line-imaging velocity interferometer system for any reflector (VISAR) measures velocities by the phase change caused by Doppler-shifted light reflecting off a moving surface, using a probe laser that propagates through transparent sample [38]. It is sensitive to the shock front planarity, as it is usually used in laser dynamic shock compression experiments [39,40]. when shock front bend angle exceeds a certain threshold, the VISAR fringe signals are lost because probe beam can not go back into VISAR system. It is about $2° - 3°$, depending on the VISAR system f value (focus distance-diameter) of lenses. Thus, a method for detecting the speed of sound by lateral unloading wave was proposed by Li et al. in transparent materals [35] . The temporal lateral release trace in X-Y space and shock velocity u_s are measured simultaneously by VISAR. When we assume that the threshold keeps q certain value when shock travels in time, the bend angle $\tan \alpha = X(t)/Y(t)$ can be obtained, where $Y(t)$ is the shock position, and $X(t)$ is edge rarefaction trace point position.

The advantage for determining speed of sound using the lateral release method is that it provides an absolute measurement for sound speed and the measurement only depends on shock velocity of shock front, spatial resolution of VISAR, and known Hugoniot $u_s - u_p$ relation.

However, it does not depend on properties from the reference material, which are required by the method from the extension to unsteady wave correction derived by Fratanduono et al. [41–43].

2.2. Experimental Configuration and Targets

The laser shock experiment was carried out at the Shenguang-III prototype laser facility (SG-IIIp), located at the Laser Fusion Research Center (CAEP) in Mianyang, China [44,45]. As a frequency-tripled Nd: glass laser, SG-IIIp facility operates at wave length of 351 nm, which usually has 8 beams with total 1200 J for each with 1–3 nanosecond (ns) phase shaped pulse ability, and 9th beam as 4000 J in 5 ns with 20 ns pulse shaping.

In this experiment, the α-quartz samples were shock compressed from X-ray indirect-drive in planar geometry. Up to 8 beams, delivering up to 800 J per beam of 351 nm UV laser in a 1–3 ns flat-top temporally shaped pulse, were focused on the 850 μm laser entrance hole (LEH) of a 30 μm thick, 3.6 mm length and 1.8 mm diameter gold hohlruam with 1 mm × 0.5 mm diagnostic hole attaching with the target packages, as shown in Figure 1a. Distributed phase plates 500 μm were used in this experiment to spatially smooth the beam profile and produce a super Gaussian intensity when projected onto the hohlruam wall. The uniform radiation field, with intense X-ray through absorption, was created in the gold hohlraum, which can deposit energy into target packages. The peak X-ray radiation temperature measured is 140–162 eV in 1–2 ns, according to the maximum pressure needed to achieve in shots. Figure 1b shows the typical X-ray radiation history in the hohlraum measured by the flat response X-ray diodes (FXRD) at 45° angle in our experiment [46], in which peak temperature achieved up to 160 eV at 1 ns in Shot 341, and 145 eV at 2 ns time in Shot 343.

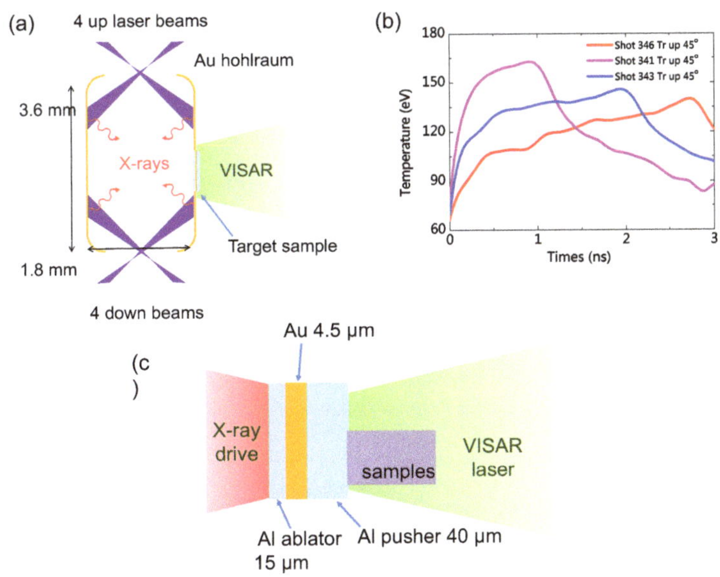

Figure 1. (**a**) Experimental setup of hohlraum-type with laser beams drive planar shock wave through X-ray. Line-imaging VISAR measures shock velocity and lateral release signal from target periphery. (**b**) Typical radiation profile of hohlraum temperature in shots from diagnostic device up 45°. (**c**) Target configuration in experiment with Al ablator, Au shield, Al pusher, and quartz sample. Free side in quartz sample was arranged in center of diagnostic hole, which is also perpendicular to Al pusher plane.

Our target packages were 1.2 mm × 1 mm flat and was consisted of a 15 μm aluminum ablator for generating shock wave by X-ray ablation, a 4.5 μm gold layer as X-ray preheat shield, and a further 35 μm aluminum layer pusher to minimize impedance mismatch with sample and the α-quartz baseplates, as shown in Figure 1c. For these experiments, 1-D hydro-dynamic simulations using the hydrocode MULTI-1D program[47] guided the peak shock pressure and temperature profile while shock propagates in target. Simulations and previous experiments also displayed negligible preheat after Al pusher for the shots at 180 eV radiation temperature in 1.5 ns. This confirms our sample was not preheated. The shock planar area from Al pusher is more than 500 μm from previous experiments' results [48–51]. Adjacent α-quartz samples were glued onto the back of Al pusher layer using an Sn thin film as epoxy glue layers < 1 μm thick. The free side of α-quartz sample is perpendicular to the Al pusher plane and it is also located in the middle of Al pusher to make sure the bent shock front is only due to the edge rarefaction from free side of sample. The α-quartz sample were >250 μm thick, with initial densities of 2.65 g/cm^3 and refractive index of 1.574 at 532 nm.

2.3. Experimental Diagnostic and Analysis

Shock velocity and lateral release in quartz were measured using a line imaging VISAR system developed in Laser Fusion Research Center (LFRC). In this experiment, the pressure in quartz is above 200 GPa, where quartz melts and forms a reflecting front [15]. The raw VISAR images, as per the figure in Figure 2, were analyzed using the Fourier transform method to determine the phase [39]. To account for 2π ambiguities in phase, two VISARs with different velocity sensitivities were used. Their sensitivities used were 5.10 and 7.30 km/s/fringe, which correspond to velocity sensitivities of 3.24 and 4.64 km/s/fringe in quartz. The VISAR images provide phase shifts resulting from changes in velocity that were analyzed using the Fourier transform method [40]. Uncertainty in determining the phase was estimated as 3% of a fringe, which resulted in <1% uncertainties due to multiple fringe jumps [40]. The VISAR system used a frequency-doubled Nd: YAG laser operating at 532 nm, and the two streak cameras used 20 ns sweep durations, resulting in temporal sensitivities of 30 ps. A sample of an extracted shock velocity profile and edge rarefaction boundary where no signal returns from the shock front is given in Figure 2.

The shock velocity u_s in the quartz was measured immediately after shock break out of Al pusher. A weak smoothing splice was applied to the velocities to mitigate the influence of random noise on the fitting routine. The first step is to select points of edge rarefaction wave trace when it incidentally enters into the shock from the side. The points were chosen within boundary of bending fringes with certain relatively intense clim. The horizontal pixels difference dX and vertical time difference dt between initial and final points in VISAR images infers the edge rarefaction wave traveling distance X and position of the shock front $Y(t)$, as $X = M \times dX$ and $Y(t) = \int_{t_0}^{t_0+dt} u_s(t)dt$, where t_0 is the time for starting point. The parameter M is the spatial resolution factor for VISAR, which is measured in static experiment before dynamical experiments. The spatial scaling $M_L = 0.536$ [μm/pixel] and $M_R = 0.64$ [μm/pixel] with certain uncertainty respectively for VISAR L and R legs in this experiment. The angle α can be obtained from the $X(t)$ and $Y(t)$ profile as $\tan\alpha = X(t)/Y(t)$.

In this work, the quartz Hugoniot data and the $u_s - u_p$ relation for α-quartz is used as $u_s = a + bu_p + cu_p^2 + du_p^3$, where $a = 6.278$, $b = 1.193$, $c = 2.505$, $d = 0.3701$. The particle velocity u_p is solved by measurement of shock velocity u_s from VISAR with uncertainty considering covariance matrix elements for $u_s - u_p$ relation by Knudson and Desjarlais [27]. We found that sound speed profile is highly dependent on the points from bent trace in VISAR images and spatial resolution of VISAR system. In our improved analysis, the average sound speed for each shot was obtained using average shock velocity between initial and final bend points, as shock density changes a little when it decays in the indirect drive experiments. This average speed of sound analysis method could mitigate the misleading information on sound speed coming from the bend trace points selecting spatial resolution 4–5 μm and random noise in VISAR images.

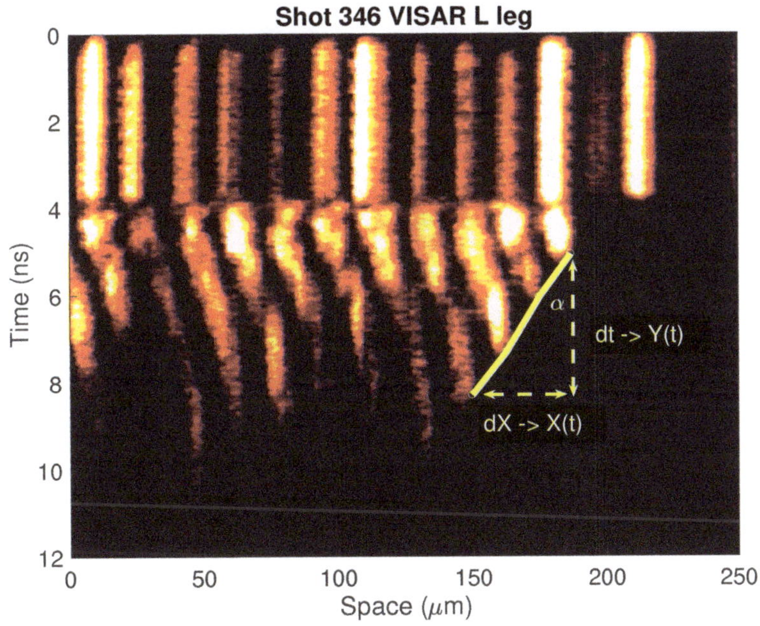

Figure 2. Raw VISAR data for shot 346. Line-imaging VISAR record provides 1-D spatial information of the shock velocity as a function of time while shock propagates through quartz sample. The edge rarefaction from free side interacts with shock wave to bend the shock wave front. The yellow line in this figure presents the trace of edge rarefaction in sample, which could be infer sound velocity with shock velocity measured from the moving fringes in VISAR images.

Uncertainties were calculated using the Monte Carlo method, which generates normally distributed random variables based on the mean and standard deviation of an observed quantity. In these experiments, independent variables in the Monte Carlo simulations were: the measured shock velocities in the quartz, coefficients in quartz Hugoniot $u_s - u_p$ relation, spatial scaling parameter M for VISAR system, spatial and temporal resolution of streak cameras. The variances in the initial density of the α-quartz is assumed to be negligible. The uncertainty in the measured velocities and coefficients in Hugoniot relation were used to determine the corresponding uncertainty in pressure, density, and particle velocity. The measured sound velocity was determined from the average of the Monte Carlo calculations for each shot. The system errors such as M parameter, VISAR spatial resolution mainly resulted in uncertainty 4–8% of sound velocities, while the random errors was estimated to be less than 1%. This level of total uncertainty is similar to techniques using laser unsteady wave method [42,43] and overtaking method with gas guns [32,33,52]. But this lateral release method provide an absolutely determined sound velocity and the uncertainties from unsteady wave method may be underestimated because they do not consider the uncertainties coming from sound velocity standard [42,43].

3. Results

The speed of sound for shocked liquid quartz was measured at pressure from 270 to 900 GPa in 7 shots experiment, as listed in Table 1. The edge rarefaction wave traces in VISAR images could generate a series of well-constrained data. The equation of state parameters for shock compressed quartz was found by the shock velocity and Hugoniot relation [27]. As mentioned previously, the speed of sound was calculated using the average velocity and the average edge rarefaction angle over the region. The uncertainty in the measurement of the sound velocity ranges from 4% to 10%.

Table 1. Experimental results on shock velocity (u_s), pressure (P), density (ρ), sound velocity (C_B), and Grüneisen parameter (γ) of shocked quartz.

Shot No.	u_s (km/s)	P (GPa)	ρ (g/cc)	C_B (km/s)	γ Parameter
340	21.63 ± 0.16	742.68 ± 11.91	6.71 ± 0.16	16.69 ± 0.66	0.76 ± 0.06
341	23.18 ± 0.16	871.28 ± 18.55	6.91 ± 0.20	15.30 ± 0.97	0.87 ± 0.07
342	19.66 ± 0.23	605.36 ± 8.12	6.43 ± 0.12	13.88 ± 0.54	0.99 ± 0.05
343	16.86 ± 0.16	430.51 ± 13.94	6.05 ± 0.06	13.94 ± 0.51	0.97 ± 0.06
344	16.68 ± 0.16	412.42 ± 4.04	6.03 ± 0.06	13.06 ± 0.51	1.07 ± 0.06
345	16.33 ± 0.16	391.78 ± 3.73	5.98 ± 0.05	14.14 ± 0.90	0.94 ± 0.11
346	13.69 ± 0.16	268.36 ± 2.29	5.64 ± 0.03	12.95 ± 1.34	1.00 ± 0.20

The data for sound velocity in silica at high pressure are summarized in Figure 3. As shown in this figure, the black squares [34] and triangles [31] represent longitudinal sound velocity data measured for crystal quartz, respectively, along quartz shock Hugoniot and spanning shock melting pressure around 200 GPa.

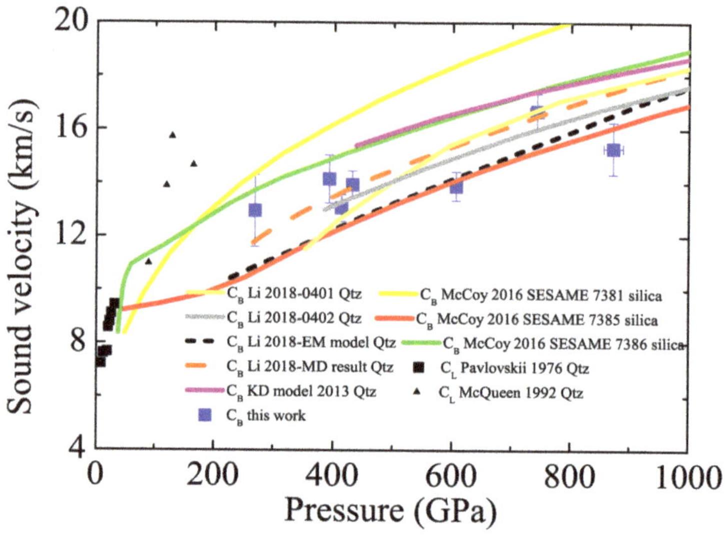

Figure 3. Measured isentropic sound velocity (blue squares with error bars) is in good agreement with 2 reported shots (light grey and yellow line) and DFT-MD calculations (orange line) in Ref. [35], as well as previous data from Pavlovskii (black squares) and McQueen results (black triangles) at pressure below 500 GPa. At higher pressure to 900 GPa, our data prefer SESAME 7385 for silica (red line) in Ref. [42] and empirical EOS model (EM model) results for quartz (black dashed line). In this figure, C_b means bulk sound velocity, C_L means longitudinal sound velocity, and *Qtz* means quartz sample.

The sound speed data from theoretical models for quartz were also plotted in Figure 3, which were calculated for initial density of fused silica as summarized in Ref. [42]. In this figure, theoretical model results from the analytic solutions (KD model) from the release isentrope [27], density functional theory molecular dynamics (DFT-MD) calculations and empirical model (EM model) from wide regime equation of state (WEOS) [35] were also presented. Most of our measured sound velocities with errors are located in the region where theoretical model data results cover the whole pressure regions from 200 GPa to 1000 GPa, but slightly lower than those data from KD model, SESAME EOS table 7381 and 7386 for silica [42]. The present sound speed results could be analyzed in two parts. At lower pressure of 200 GPa to 500 GPa, the sound velocity agrees with decay-laser shock results and DFT-MD results in Ref. [35]. But as the pressure goes above 600 GPa, our sound velocity data indicate a gentle increase in slope at high pressures from the whole data, compared with the previous theoretical and experimental results.

Liquid silica at extremely high shock compression undergo a significant dissociation transition where SiO_2 dissociates atomic Si and O. The broad peak in specific heat and $u_s - u_p$ Hugoniot curve observed in experiments were determined for indicating short-range structure or the energetics of bond breaking [24,27]. The bulk sound velocity behavior in 700–900 GPa may infer this bonded to atom liquid phase transition in quartz. A smooth increase in present sound velocity data is shown in this figure, although some are beyond the errors, and more experimental data for shocked quartz around pressure 700–850 GPa are still lacking, which could help to show more clear changes without being misled by the experimental uncertainties of data.

The Grüneisen parameter γ, defined by $\gamma = V(\frac{dP}{dE})_V$, is a fundamental thermodynamic quantity with wide applications. For liquid silica, it affects the adiabatic temperature gradient and the amount of heat conducted along the adiabatic. It was assumed to be a constant by different model by DFT-MD calculations [27]. γ can be determined from the slope of Hugoniot data and the isentrope at a point in phase space, which is related as

$$\gamma = \frac{2(V(\frac{dP}{dV})_H + \frac{C_s^2}{V})}{P_H + (\frac{dP}{dV})_H(V_0 - V_H)}, \tag{2}$$

where $(dP/d\rho)_H$ is the slope of Hugoniot curve at volume V_H and pressure P, subscript 0 indicates zero-pressure conditions and C_s is the isentropic sound velocity at Hugoniot state.

In this work, γ parameter for liquid quartz at high pressure was calculated using Equation (2) for each shot. The uncertainty in Grüneisen parameter was also calculated using the Monte Carlo method to be 5% to 20%, which depends on the pressure. The slope $(dP/d\rho)_H$ is adopted by Rankine–Hugoniot relationship in Ref. [27]. The increasing uncertainty in the slope of Hugoniot at high pressures also increases that in γ parameter.

Figure 4 presents our data and previous data from experimental and theoretical models. Most of results at higher pressure fall into the region bounded by $V\Delta P/\Delta E$ average values (light blue line) and DFT-MD results (blue line) in Ref. [35]. The Grüneisen parameter decreases with shock temperature and pressure over the experimental range in general view. But they also show a slower slope comparing with that of experimental results (0401 black line and 0402 red line) and DFT-MD calculations. They also do not favor any SESAME table, which predicted a significantly stiffer result. For the densities above 6.0 g/cm^3, experimental data approach 0.7–0.8. This is in agreement with previous results that implied the validity of a constant γ parameter for liquid silica above high pressure 400 GPa. But this value is bigger compared with the constant (the value for the ideal gas.) $\gamma = 0.66$ (black dashed line) in Refs. [27,42]. The data in this work also show a slower decay in γ parameter, which is directly relevant to the sound velocity behavior at high pressure. Although more data are still needed to make up around the regions in future experiments, they may infer the boned-atomic dissociation phase transition in liquid quartz, as γ parameter directly relates to the pressure and energy, which is required to break chemical bonds for dissociation of liquid SiO_2 molecule into Si and O atom.

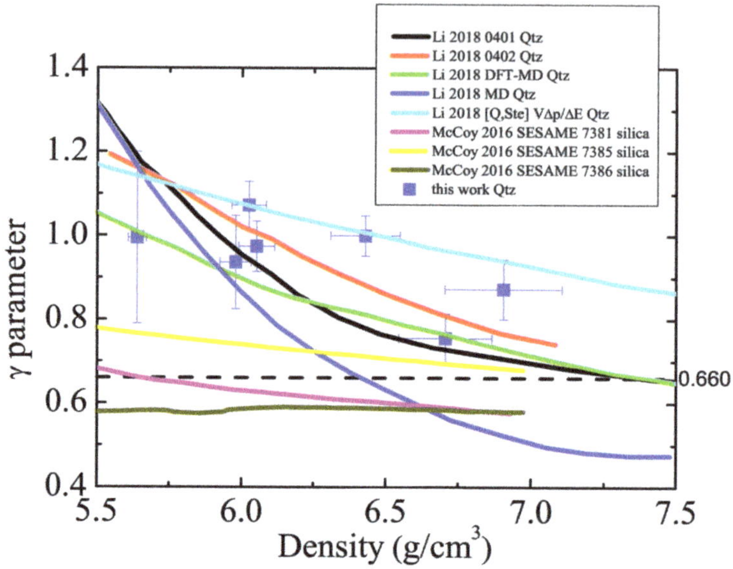

Figure 4. Measurements of Grüneisen parameter in quartz (blue square with error bars). Most of present results fall into region bounded by average values (light blue line) and DFT-MD results (blue line) in Ref. [35]. All data do not favor any SESAME table Ref. [42] and are greater than the theoretical limit $\gamma = 0.66$. The Grüneisen parameter measured is in agreement with previous experimental data in 0401 (black line) and 0402 (blue line), as density is below 6.2 g/cm^3. The slower decay process in the Grüneisen parameter was found with consideration of the errors as the density goes from 6.5 to 7.0 g/cm^3.

4. Conclusions

The sound velocity of shocked α-quartz was measured using lateral release wave from the periphery of the crystal in the regions of 250 to 900 GPa, where it is melted. The improved analysis method in the present study demonstrates lateral release method for directly measuring the sound velocity by VISAR, proposed by Li et al. [35], when Rankine–Hugoniot relation is known. The present results are consistent with previous experimental results and predictions by DFT-MD. The sound velocity in liquid quartz above 600 GPa shows more gentle slope than previous experimental results and prefers SEASAME EOS 7385 for silica [42] and the empirical model [35]. The uncertainties by Monte Carlo calculations are similar to those of the previous results with uncertainties. The Grüneisen parameter was also calculated from sound velocity and quartz Hugoniot data, and the result decreases with increasing shock pressure. The results agree with the theoretical models as a shallow slope relative to the pressure. The behavior for measured sound velocity and γ parameter at extreme high pressures may be directly related to the boned to atomic phase transition found in shocked liquid silica at high pressure, although more experimental data should be provided in future experiments.

The sound speed revisited in our experiment provide further information on dissociation and metallization for liquid quartz at extreme conditions. The more gentle behavior we observed in present experiments may give us a new vision about transport properties of high-pressure liquid silica, although it requires further study. Our agreements with previous results increase our confidence in our theoretical model in Super-Earth interior conditions and help our understanding of the properties of magma oceans and melting on exoplanets.

Author Contributions: Project lead, T.S., L.S.; experiment participants, W.Y., Y.Z., Z.G., Y.Z., X.D.; formal analysis, Y.Z.; investigation, L.S., Z.W.; resources, Z.W., J.Y.; data analysis, L.S., H.Z.; writing—original draft preparation, L.S. and H.Z.; writing, review, and editing, L.S., Y.Z. and T.S. All authors have read and agreed to the published version of the manuscript.

Funding: This work was supported by the National Key R&D Program of China under Contract No. 2017YFA0403201 and the National Natural Science Foundation of China No. 12074351.

Data Availability Statement: All VISAR results and sample characterization data are available upon request.

Acknowledgments: The authors would like to thank the technical team members, Zhi Jun Li, Sai Jin, Run Chang Zhao, Meng Shen Yang, Tai Ming Yi, and Wei Chao Tong at LFRC, for their invaluable work producing the targets and operating the facility. Liang Sun would like to thank Shao Yong Tu at LFRC for helpful discussion on data analysis. The authors also want to thank Hua Shu for helpful discussion on physical analysis.

Conflicts of Interest: The authors declare no conflict of interest.

References

1. Duffy, T.; Madhusudhan, N.; Lee, K.K.M. Mineralogy of Super-Earth Planets. In *Treatise on Geophysics*; Schubert, G., Ed.; Elsevier: Amsterdam, The Netherlands, 2015; pp. 393–416.
2. Rauer, H.; Catala, C.; Aerts, C.; Appourchaux, T.; Benz, W.; Brandeker, A.; Christensen-Dalsgaard, J.; Deleuil, M.; Gizon, L.; Goupil, J.; et al. The PLATO 2.0 mission. *Exp. Astron.* **2014**, *38*, 249–330. [CrossRef]
3. Lissauer, J.; Dawson, R.I.; Tremaine, S. Advances in exoplanet science from Kepler. *Nature* **2014**, *513*, 336–344. [CrossRef]
4. Valencia, D.; Connel, R.J.O.; Sasselov, D. Internal structure of massive terrestrial planets. *Icarus* **2006**, *181*, 545–554. [CrossRef]
5. Swift, D.C.; Eggert, J.H.; Hicks, D.C.; Hamel, S.; Caspersen, K.; Schwegler, E.; Collins, G.W.; Nettelmann, N.; Ackland, G.J. Mass-radius relationships for exoplanets. *Astrophys. J.* **2012**, *744*, 59. [CrossRef]
6. Zeng, L.; Sasselov, D.D.; Jacobsen, S.B. Mass-radius relation for rocky planets based on Prem. *Astrophys. J.* **2016**, *819*, 127. [CrossRef]
7. Unterbornal, C.T.; Dismukes, E.E.; Panero, W.R. Scaling the Earth: A sensitivity analysis of terrestrial exoplanetary interior models. *Astrophys. J.* **2016**, *819*, 32. [CrossRef]
8. Olson, P. Mantle control of the geodynamo: Consequences of top-down regulation. *Geochem. Geophys. Geosyst.* **2006**, *17*, 1935–1956. [CrossRef]
9. Lammer, H. What makes a planet habitable? *Astron. Astrophys. Rev.* **2009**, *17*, 181–249. [CrossRef]
10. Soubiran F.; Militzer B. Electrical conductivity and magnetic dynamos in magma oceans of Super-Earths. *Nat. Comm.* **2018**, *9*, 3883. [CrossRef]
11. Stixrude, L. Melting in Super-Earths. *Philos. Trans. R. Soc. A* **2014**, *372*, 20130076. [CrossRef]
12. Hemley, R.; Prewitt, C.T.; Kingma, K.J. High-pressure behavior of silica. In *Silica: Physical Behavior, Geochemistry and Materials Applications*; Manghnani, P.J., Prewitt, C.T., Gibbs, G.V., Eds.; Mineralogical Society of America: Blacksburg, Virginia, 1994; Volume 29, pp. 41–81.
13. Wentzcovitch, R.M.; Silva, C.D.; Chelikowsky, J.R.; Binggeli, N. A new phase and pressure induced amorphization in silica. *Phys. Rev. Lett.* **1998**, *80*, 2149–2152. [CrossRef]
14. Haines, J.; Leger, J.M.; Gorelli, F.; Hanfland, M. Crystalline post-quartz phase in silica at high pressure. *Phys. Rev. Lett.* **2001**, *87*, 155503. [CrossRef] [PubMed]
15. Akins, J.A.; Ahrens, T.J. Dynamic compression of SiO_2: A new interpretation. *Geophys. Res. Lett.* **2002**, *29*, 31-1–31-4. [CrossRef]
16. Luo, S.-N.; Akins, J.A.; Ahrens, T.J.; Asimov, P.D. Shock-compressed $MgSiO_3$ glass, enstatite, olivine, and quartz: Optical emission, temperatures, and melting. *J. Geophys. Rev.* **2004**, *109*, B05205. [CrossRef]
17. Dubrovinsky, L.S.; Dubrovinskaia, N.A.; Prakapenka, V.; Seifert, F.; Langenhorst, F.; Dmitriev, V.; Weber, H.-P.; Bihan, T.L. A class of new high-pressure silica polymorphs. *Phys. Earth Planet. Int.* **2004**, *143–144*, 231–240. [CrossRef]
18. Hu, Q.Y.; Shu, J.-F.; Yang, W. G.; Park, C.; Chen, M. W.; Fujita, T.; Mao, H.-K.; Sheng, H. W. Stability limits and transformation pathways of α-quartz under high pressure. *Phys. Rev. B* **2017**, *95*, 104112. [CrossRef]
19. Tracy, S.J.; Turneaure, S.J.; Duffy, T.S. In situ X-Ray Diffraction of Shock-Compressed Fused Silica. *Phys. Rev. Lett.* **2018**, *120*, 137702. [CrossRef]
20. Prescher, C.J.; Prakapenka, V.B.; Stefanski, J.; Jahn, S.; Skinner, L.B.; Wang, Y.-B. Beyond sixfold coordinated Si in SiO_2 glass at ultrahigh pressures. *Proc. Natl. Acad. Sci. USA* **2017**, *114*, 10041–10046. [CrossRef] [PubMed]
21. Asimov, P.D. Dynamic compression. In *Treatise on Geophysics*; Schubert, G., Ed.; Elsevier: Amsterdam, The Netherlands, 2015; pp. 393–416.
22. Hicks, D.G.; Boehly, T.R.; Celliers, P.M.; Eggert, J.H.; Vianello, E.; Meyerhofer, D.D.; Collins, G.W. Shock compression of quartz in the high-pressure fluid regime. *Phys. Plasmas* **2005**, *12*, 082702. [CrossRef]

23. Knudson, M.D.; Desjarlais, M.P. Shock Compression of Quartz to 1.6 TPa: Redefining a Pressure Standard. *Phys. Rev. Lett.* **2009**, *103*, 225501. [CrossRef]
24. Hicks, D.G.; Beohly, T.R.; Eggert, J.H.; Miller, J.E.; Celliers, P.M.; Collins, G.W. Dissociation of Liquid Silica at High Pressures and Temperatures. *Phys. Rev. Lett.* **2006**, *97*, 025502. [CrossRef] [PubMed]
25. Qi, T.; Millot, M.; Kraus, R.G.; Root, S.; Hamel, S. Optical and transport properties of dense liquid silica. *Phys. Plasmas* **2015**, *22*, 062706. [CrossRef]
26. Boehly, T.R.; Miller, J.E.; Meyerhofer, D.D.; Eggert, J.H.; Celliers, P.M.; Hicks, D.G.; Collins, G.W. *Shock Compression of Conducted Matter-2007*; Elert, M., Furnish, M.D., Chau, R., Holmes, N., Nguyen, J., Eds.; American Institute of Physics: Melville, NY, USA, 2007; Volume 955, p. 19.
27. Knudson, M.D.; Desjarlais, M.P. Adiabatic release measurements in α-quartz between 300 and 1200 GPa: Characterization of α-quartz as a shock standard in the multimegabar regime. *Phys. Rev. B* **2013**, *88*, 184107. [CrossRef]
28. Kraus, R.G.; Stewart, S.T.; Swift, D.C.; Bolme, C.A.; Smith, R.F.; Hamel, S.; Hammel, B.D.; Spaulding, D.K.; Hicks, D.G.; Eggert, J.H.; et al. Shock vaporization of silica and the thermodynamics of planetary impact events: Shock vaporization of silica. *J. Geophys. Res.* **2012**, *117*, E09009. [CrossRef]
29. Millot, M.; Dubrovinskaia, N.; Cernok, A.; Blaha, S.; Dubrovinsky, L.; Braun, D.G.; Celliers, P.M.; Collins, G.W.; Eggert, J. H.; Jeanloz, R. Shock compression of stishovite and melting of silica at planetary interior conditions. *Science* **2015**, *347*, 418 [CrossRef]
30. Denoeudn, A.; Benuzzi-Mounaix, A.; Ravasio, A.; Dorchies, F.; Leguay, P.M.; Gaudin, J.; Guyot, F.; Brambrink, E.; Koenig, M.; Le Pape, S.; et al. Metallization of Warm Dense SiO_2 Studied by XANES Spectroscopy. *Phys. Rev. Lett.* **2014**, *113*, 116404. [CrossRef]
31. Pavlovskii, M.N. Measurements of the velocity of sound in shock-compressed quartzite, dolomite, anhydrite, sodium chloride, paraffin, plexiglas, polyethylene, and fluoroplast-4. *J. Appl. Mech. Tech. Phys.* **1976**, *17*, 709. [CrossRef]
32. Duffy, T.S.; Ahrens, T.J. Sound velocities at High pressure and Temperature and Their Geophysical Implications. *J. Geophys. Res.* **1992**, *97*, 4503–4520. [CrossRef]
33. Tan, Y.; Yu, Y.Y.; Dai, C.-D.; Jin, K.; Wang, Q.S.; Hu, J.-B.; Tan, H. Hugoniot and sound velocity measurements of bismuth in the range of 11–70 GPa. *J. Appl. Phys.* **2013**, *113*, 093509. [CrossRef]
34. McQueen, R.G. The Velocity of Sound behind Strong Shocks in SiO_2. In *Shock Compression of Condensed Matter-1991*; Dick, R.D., Forbes, J.W., Tasker, D.G., Eds.; Elsevier: Amsterdam, The Netherlands, 1992; p. 75.
35. Li, M.; Zhang, S.; Zhang, H.P.; Zhang, G.M.; Wang, F.; Zhao, J.H.; Sun, C.W.; Jeanloz, R. Continuous Sound Velocity Measurements along the Shock Hugoniot Curve of Quartz. *Phys. Rev. Lett.* **2018**, *120*, 215703. [CrossRef]
36. Altshuler, V.L.; Kormer, S.B.; Brazhnik, M.I.; Vladimirov, L.A.; Speranskaya, M.P.; Funtikov, A.I. The isentropic compression of aluminum, copper, lead and iron at high pressures. *Sov. Phys. JETP* **1960**, *11*, 766.
37. Zeldovich, Y.B.; Raizer, Y.P. *Physics of Shock Waves and High Temperature Hydrodynamic Phenomena*; Hayes, W.D., Probstein, R.F., Eds.; Dover Publications, Inc.: Mineola, NY, USA, 2002.
38. Barker, L.M.; Hollenbach, R.E. Laser interferometer for measuring high velocities of any reflecting surface. *J. Appl. Phys.***1972**, *43*, 4669. [CrossRef]
39. Celliers, P.M.; Collins, G.W.; Da Silva, L.B.; Gold, D.M.; Cauble, R. Accurate measurement of laser-driven shock trajectories with velocity interferometry. *Appl. Phys. Lett.* **1998**, *73*, 1320. [CrossRef]
40. Celliers, P.M.; Bradley, D.K.; Collins, G.W.; Hicks, D.G.; Boehly, T.R.; Armstrong, W.J. Line-imaging velocimeter for shock diagnostics at the OMEGA laser facility. *Rev. Sci. Instrum.* **2004**, *75*, 4916. [CrossRef]
41. Fratanduono, D.E.; Munro, D.H.; Celliers, P.M.; Collins, G.W. Hugoniot experiments with unsteady waves. *J. Appl. Phys.* **2014**, *116*, 033517. [CrossRef]
42. McCoy, C.A.; Gregor, M.C.; Polsin, M.C.; Fratanduono, D.E.; Celliers, P.M.; Boehly, T.R.; Meyerhofer, D.D. Measures of the sound velocity of shock-compressed liquid silica to 1100 GPa. *J. Appl. Phys.* **2016**, *120*, 235901. [CrossRef]
43. McCoy, C.A.; Hu, S.X.; Marshall, M.C.; Polsin, M.C.; Fratanduono, D.E.; Ding, Y.H. Celliers, P.M.; Boehly, T.R.; Meyerhofer, D.D. Measurement of the sound velocity and Grüneisen parameter of polystyrene at inertial confinement fusion conditions. *Phys. Rev. B* **2020**, *102*, 184102. [CrossRef]
44. Zheng, W.-G.; Wei, X.-F.; Zhu, Q.-H.; Jing, F.; Hu, D.-X.; Su, J.-Q.; Zheng, K.-X.; Yuan, X.-D.; Zhou, H.; Dai, W.-J.; et al. Laser performance of the SG-III laser facility. *High Power Laser Sci. Eng.* **2016**, *4*, e21. [CrossRef]
45. Gong, T.; Hao, L.; Li, Z.-C.; Yang, D.; Li, S.-W.; Guo, L.; Zou, S.-Y.; Liu, Y.-Y.; Jiang, X.-H.; Peng, X.-S.; et al. Recent research progress of laser plasma interactions in Shenguang laser facilities. *Matter Radiat. Extremes.* **2019**, *4*, 055202. [CrossRef]
46. Li, Z.-C.; Jiang, X.-H.; Liu, S.-Y.; Huang, T.-X.; Zheng, J.; Yang, J.-M.; Li, S.-W.; Guo, L.; Zhao, X.-F.; Du, H.-B.; et al. A novel flat-response X-ray detector in the photon energy range of 0.1–4 keV. *Rev. Sci. Instrum.* **2010**, *81*, 073504. [CrossRef]
47. Ramis, R.; Meyer-ter-Vehn, J. MULTI-IFE–A one-dimensional computer code for Inertial Fusion Energy (IFE) target simulations, *Comput. Phys. Commun.* **2016**, *203*, 226–237. [CrossRef]
48. Liu, W.; Duan, X.-X.; Jiang, S.-E.; Wang, Z.-B.; Sun, L.; Liu, H.; Yang, W.-M.; Zhang, H.; Ye, Q.; Li, Y.-Y.; et al. Laser-driven shock compression of gold foam in the terapascal pressure range. *Phys. Plasmas.* **2018**, *25*, 062707. [CrossRef]
49. Hartley, N. J.; Zhang, C.; Duan, X.-X.; Huang, L.-G.; Jiang, S.-E.; Li, Y.; Yang, L.; Pelka, A.; Wang, Z.-B.; Yang, J.-M.; et al. Dynamically pre-compressed hydrocarbons studied by self-impedance mismatch. *Matter Radiat. Extrem.* **2020**, *5*, 028401. [CrossRef]

50. Duan, X.-X.; Zhang, C.; Guan, Z.-Y.; Sun, L.; Peng, X.-S.; Liu, H.; Yang, W.-M.; Li, Y.-L.; Zhang, H.; Ye, Q.; et al. Transparency measurement of lithium fluoride under laser-driven accelerating shock loading. *J. Appl. Phys.* **2020**, *128*, 015902. [CrossRef]
51. Wang, P.; Zhang, C.; Jiang, S.-E.; Duan, X.-X.; Zhang, H.; Li, L.-L.; Yang, W.-M.; Liu, Y.-G.; Li, Y.-L.; Sun, L.; et al. Density-dependent shock Hugoniot of polycrystalline diamond at pressures relevant to ICF. *Matter Radiat. Extrem.* **2021**, *6*, 033902. [CrossRef]
52. Berryman, E.J.; Winey, J.M.; Gupta, Y.M.; Duffy, T.S. Sound Velocities in Shock-Synthesized Stishovite to 72 GPa. *Geophys. Res. Lett.* **2019**, *46*, 13,697–13.703. [CrossRef]

minerals

MDPI

Article

Self-Consistent Thermodynamic Parameters of Diopside at High Temperatures and High Pressures: Implications for the Adiabatic Geotherm of an Eclogitic Upper Mantle

Chang Su [1,2,3], Dawei Fan [2], Jiyi Jiang [1,3], Zhenjun Sun [1,3], Yonggang Liu [2,*], Wei Song [2], Yongge Wan [1,3], Guang Yang [4] and Wuxueying Qiu [1]

1 School of Earth Sciences, Institute of Disaster Prevention, Sanhe 065201, China; suchang@cidp.edu.cn (C.S.); jiangjiyi@cidp.edu.cn (J.J.); sunzhenjun@cidp.edu.cn (Z.S.); wanyg217217@vip.sina.com (Y.W.); qiuwuxueying@outlook.com (W.Q.)
2 Key Laboratory of High Temperature and High Pressure Study of the Earth's Interior, Institute of Geochemistry, Chinese Academy of Sciences, Guiyang 550081, China; fandawei@vip.gyig.ac.cn (D.F.); songwei@vip.gyig.ac.cn (W.S.)
3 Hebei Key Laboratory of Earthquake Dynamics, Sanhe 065201, China
4 School of Information Engineering, Institute of Disaster Prevention, Sanhe 065201, China; yangguang@cidp.edu.cn
* Correspondence: liuyonggang@vip.gyig.ac.cn

Citation: Su, C.; Fan, D.; Jiang, J.; Sun, Z.; Liu, Y.; Song, W.; Wan, Y.; Yang, G.; Qiu, W. Self-Consistent Thermodynamic Parameters of Diopside at High Temperatures and High Pressures: Implications for the Adiabatic Geotherm of an Eclogitic Upper Mantle. *Minerals* **2021**, *11*, 1322. https://doi.org/10.3390/min11121322

Academic Editors: Lidong Dai, Haiying Hu, Jianjun Jiang and Leonid Dubrovinsky

Received: 22 October 2021
Accepted: 22 November 2021
Published: 26 November 2021

Abstract: Using an iterative numerical approach, we have obtained the self-consistent thermal expansion, heat capacity, and Grüneisen parameters of diopside ($MgCaSi_2O_6$) over wide pressure and temperature ranges based on experimental data from the literature. Our results agree well with the published experimental and theoretical data. The determined thermodynamic parameters exhibit nonlinear dependences with increasing pressure. Compared with other minerals in the upper mantle, we found that the adiabatic temperature gradient obtained using the thermodynamic data of diopside is larger than that of garnet while lower than that of olivine, when ignoring the Fe incorporation. Combining our results with thermodynamic parameters of garnet obtained in previous studies, we have estimated the adiabatic temperature gradient and geotherm of an eclogitic upper mantle in a depth range of 200–450 km. The results show that the estimated adiabatic temperature gradient of the eclogite model is ~16% and ~3% lower than that of the pyrolite model at a depth of 200 km and 410 km, respectively. However, the high mantle potential temperature of the eclogite model leads to a similar temperature as the pyrolite model in a depth range of 200–410 km.

Keywords: diopside; thermodynamic properties; eclogite model; adiabatic temperature gradient; adiabatic geotherm

1. Introduction

Temperature is one of the fundamental quantities for determining the physical and chemical properties of the Earth's interior. At present, however, it is impossible to measure the temperature of the deep Earth directly. Since the seismic discontinuities in the mantle are usually related to mineral phase transitions [1], the temperatures at the seismic discontinuities (i.e., the 410 km, 520 km, 660 km discontinuities and the mantle–core boundary) can be determined by comparing the depth of the seismic velocity discontinuities with the phase transition pressures of mantle minerals [2]. In addition, since the radiative transfer of heat is regarded to be negligible in the mantle, the temperature gradient of the mantle is expected to be nearly adiabatic [3]. As a consequence, the adiabatic temperature gradient ($(\partial T/\partial z)_S$) between two seismic discontinuities can be estimated using thermodynamic parameters at high temperatures and high pressures (P–Ts) via Equation (1) [4,5]:

$$\left(\frac{\partial T}{\partial z}\right)_S = \frac{\alpha g T}{C_P} = \left(\frac{\gamma T}{K_S}\right)\frac{g}{V} \quad (1)$$

where T and z refer to the temperature and depth, respectively; g refers to the gravitational acceleration; K_S refers to the adiabatic bulk modulus; V refers to the specific volume; and α, C_P, and γ refer to the thermal expansion, isobaric heat capacity, and Grüneisen parameters, respectively.

Based on thermodynamic principles and a model of the Earth [6], Stacey provided us with important insights into the thermal state of the Earth's interior [7]. Then, considering the temperatures at the seismic discontinuities, several authors also derived the thermodynamic parameters of the deep Earth with increasing depth using various thermal equations of state (EoS) [8–10]. In these studies, different seismic models and empirical equations or coefficients used in the estimations would cause significant discrepancies in the derived geotherms, consequently influencing the investigations into the physical and chemical properties of the mantle.

By comparing the elastic properties of major mantle minerals at *P*–*T*s to global seismic models, the pyrolite and eclogite mantle models were proposed, which are particularly important to our understanding of deep-Earth mineral composition [1]. The pyrolite model is mainly composed of ~60% olivine, ~15% garnet, ~15% clinopyroxene and ~10% orthopyroxene at a depth of ~200 km, representing the bulk mantle mineral assemblages [11]. The eclogite model mainly consists of ~30% garnet and ~70% clinopyroxene at a depth of 200 km, which refers to the bulk mineral assemblage of the subducted oceanic crust [12]. A decade ago, based on the thermal equations of states (EoS) of forsterite (Mg_2SiO_4) and its high-pressure phase minerals [13–17], Katsura et al. [18] presented the adiabatic temperature gradient and geotherm of a pyrolitic mantle by using Equation (1). This work presented the idea that the adiabatic geotherm can be estimated using the thermodynamic properties of mantle minerals and specifically implied that the calculated adiabatic geotherm may be associated with the abundance of the minerals in the mantle. Recently, Su et al. modified a numerical iterative procedure and obtained the self-consistent thermodynamic properties of Fe-bearing olivine and garnet end-members at the *P*–*T* range of the upper mantle [19–21] and presented the adiabatic temperature gradients by using Equation (1). The results indicated that not only different minerals but also chemical composition variations in the same groups of minerals might affect the estimated adiabatic temperature gradient. Thus, accurate thermodynamic parameters of mantle minerals at simultaneous *P*–*T*s are urgently required to provide a better understanding of the relationship between the geotherm and the mineral composition.

Clinopyroxene is one of the main constituents of the upper mantle and the lower crust, and diopside ($CaMgSi_2O_6$) with a space group of *C2/c* is a vital calcium end-member of Ca-rich clinopyroxene [22]. Compared to the plentiful data on its elastic properties [23,24], thermodynamic data on diopside, especially under high-pressure conditions, are still lacking. Theoretically, the high-temperature and high-pressure thermal expansion of diopside can be derived from the volume at *P*–*T*s measured through X-ray diffraction experiments [23,25]. However, the deduction of the pressure–volume–temperature (*P*-*V*-*T*) relation requires empirical EoSs such as the third order Birch–Murnaghan EoS [26] or the Mie–Grüneisen–Debye EoS [27]. To date, the thermal expansion of diopside has only been published for ambient pressure conditions [28,29]. Meanwhile, since the heat capacity of diopside has only been measured at ambient pressure conditions [30–32], the available Grüneisen parameter data of diopside are consequently limited at ambient pressure conditions as well [33]. Moreover, the elastic and thermodynamic parameters of diopside have tended to be obtained individually via different methods. Still, it has been demonstrated that great care must be taken to preserve thermodynamic self-consistency in describing the temperature and pressure dependence of a material simultaneously [34,35].

This study reports the unit-cell volume, elastic moduli, thermal expansion, heat capacity, and Grüneisen parameters of diopside at high temperatures and high pressures, using an iterative numerical approach. To investigate the effect of mineral composition on the adiabatic temperature gradient, we present adiabatic temperature gradients calculated using thermodynamic data on olivine, garnet, and diopside. Finally, we apply the obtained thermodynamic data on diopside to estimate the adiabatic temperature gradient and geotherm of an eclogitic upper mantle in a depth range of 200–450 km, based on the assemblage of eclogitic mineral compositions, and compare our predictions with previous studies.

2. Methods

2.1. Theory

The approach used in this study has been described in published papers [19–21]; the fundamental equations are shown below. Note that the uncertainties of the derived parameters are estimated from the experimental measuring error propagations using a Taylor series expansion [20].

The thermal expansion at constant pressure is related to volume (V) as $\alpha = (1/V) \times (\partial V/\partial T)_P$; the integration of this expression yields [36]

$$V(T) = V_{T0} exp\left[\int_{T_0}^{T} \alpha(T)dT\right] \tag{2}$$

where V_{T0} is the volume at a reference temperature T_0. The isothermal derivative of V with respect to pressure can be written as

$$\left(\frac{\partial V}{\partial P}\right)_T = -V^2\left(\frac{1}{v_B^2} + \frac{T\alpha^2}{C_P}\right) \tag{3}$$

where v_B refers to the bulk velocity, which is related to the P-wave velocity (v_P) and S-wave velocity (v_S) as

$$v_B = \left(v_p^2 - \frac{4}{3}v_s^2\right)^{\frac{1}{2}} \tag{4}$$

Moreover, the isothermal derivative of the heat capacity with respect to pressure can be evaluated by

$$\left(\frac{\partial C_P}{\partial P}\right)_T = -VT\left[\alpha^2 + \left(\frac{\partial \alpha}{\partial T}\right)_P\right] \tag{5}$$

To start the calculation, first, we use the experimental volume data at high temperatures to obtain $\alpha(P_0,T)$ at ambient pressure by Equation (2). With $\alpha(P_0,T)$ and the relationship between C_P and temperature at ambient pressure $C_P(P_0,T)$, the approximate V at an arbitrary reference pressure can be estimated using Equation (3). Then, the resulting V is used to update the value of α and C_P at the same pressure as in Equations (2) and (5), respectively. Hence, iterations of this loop lead to converged values of V, α, and C_P as a function of temperature and pressure based on experimental elastic wave velocity. Furthermore, the self-consistent elastic parameters, including the adiabatic bulk modulus (K_S), isothermal bulk modulus (K_T), and shear modulus (G), as well as the Grüneisen parameter (γ), can be consequently derived (Equations (6)–(9)).

$$K_S = \frac{1}{V}\left(v_p^2 - \frac{4}{3}v_s^2\right) \tag{6}$$

$$K_T = \frac{K_S}{1 + \alpha\gamma T} \tag{7}$$

$$G = \frac{1}{V}v_S^2 \tag{8}$$

$$\gamma = \frac{\alpha K_S V}{C_P} \tag{9}$$

2.2. Thermoelastic Data

2.2.1. Elastic Wave Velocity under High-Temperature and High-Pressure Conditions

The elastic wave velocities of diopside have been obtained using Brillouin spectroscopy [37], ultrasonic interferometry [23], and first-principle calculations [24] under high-pressure conditions. The experimental data presented by Li and Neuville [23] indicated that v_P and v_S of diopside increase linearly with pressure and temperature within their measured *P*–*T* range. Meanwhile, the results by Sang and Bass [37] and Zou et al. [24] suggested that both v_P and v_S show noticeable nonlinear dependence on pressure over ~8 GPa. Hence, we use the data from Li and Neuville [23] and fit it with a two-dimensional equation:

$$v = m_0 + m_1 \times (T - 300) + m_2 \times P + m_3 \times P^2 \quad (10)$$

The fitting coefficients are listed in Table 1, with *P* in GPa, *T* in K, and v_P and v_S in m/s. The adjusted R^2 values are 0.979 and 0.953 for v_P and v_S, respectively. The calculated elastic wave velocities for Equation (10) agree well with the available data [24,37]. The largest differences are less than 0.5% and 1.2% for v_P and v_S, respectively, within our calculated *P*–*T* range.

Table 1. Fitting coefficients for Equation (10).

	m_0	m_1	m_2	m_3
v_P	8061.78 (±13.41)	−0.402 (±0.019)	101.21 (±4.42)	−1.771 (±0.317)
v_S	4712.23 (±13.78)	−0.329 (±0.019)	33.60 (±4.55)	−0.388 (±0.325)

2.2.2. Unit-Cell Volume under High-Temperature and Ambient Pressure Conditions

The ambient pressure unit-cell volumes of diopside were measured using X-ray diffraction at high temperatures [28,29,38,39]. In this study, we fitted the experimental volume data from Pandolfo et al. [28] to Equation (2) with the program EosFit7 GUI [40], thereby obtaining the thermal expansion as a function of temperature using the formula suggested by Fei [41]:

$$\alpha(T) = 2.8485(12) \times 10^{-5} + 4.473(11) \times 10^{-9}T - 0.2251(12)T^{-2} \quad (11)$$

Here, we take T_0 = 303 K and V_{T0} = 428.70 Å^3, which are the values suggested by Pandolfo et al. [28]. The volumes calculated using Equations (2) and (11) show good agreement with the previous experimental results [28,29,38,39], with the largest difference being less than 0.3%.

2.2.3. Heat Capacity under High-Temperature and Ambient Pressure Conditions

The heat capacity of diopside at high temperatures below 1000 K has been determined by Stebbins et al. [32] and Krupka et al. [30] by using calorimetry. Then, Richet and Fiquet [31] extended the measuring temperature to ~1600 K, and the experimental data were fitted to different forms. Here, we use Equation (12), which was suggested by Richet and Fiquet [31] to be the best for describing their data:

$$C_P = -186.908 + 57.729\ln(T) + 5.5483 \times 10^4 T^{-1} - 2.2557 \times 10^7 T^{-2} + 2.4436 \times 10^9 T^{-3} \quad (12)$$

where C_P is in J/mol K and *T* is in K. The quality of the fit is excellent, with an adjusted R^2 value of 0.993 and the largest error being 0.12%.

3. Results

3.1. Unit-Cell Volumes at High Temperatures and High Pressures

Figure 1 presents comparisons between our calculated unit-cell volume of diopside (Supplementary Materials) and previously published results. At 300 K, our data shows good agreement with the results measured using X-ray diffraction [42,43]. The excellent consistency with available experimental results also holds at high temperatures up to 1073 K [23,25], with the largest differences being less than ~0.3%. Though our unit-cell volume is slightly lower than the first-principle calculation result under high-temperature conditions [24], the uncertainty is within 1.2%.

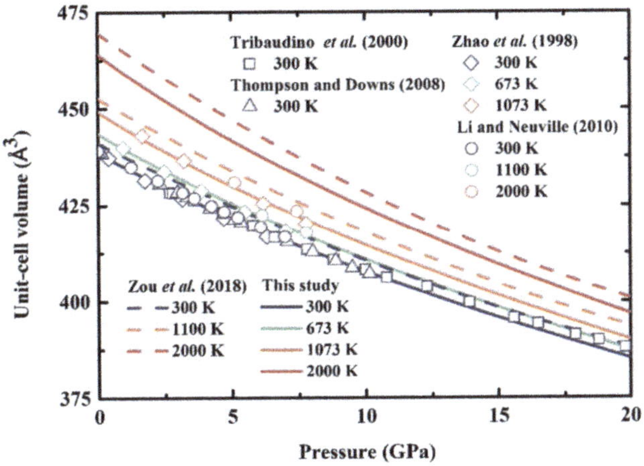

Figure 1. Unit-cell volume of diopside as a function of pressure at various temperatures. Solid lines refer to our calculated results, and dashed lines refer to the first-principle calculation results from Zou et al. (2018) [24]. The squares and triangles represent the measurements at 300 K by Tribaudino et al. (2000) [43] and Thompson and Downs (2008) [42], respectively. The diamonds and circles correspond to the measurements at various temperatures by Zhao et al. (1998) [25] and Li and Neuville (2010) [23], respectively.

3.2. Elastic Properties at High Temperatures and High Pressures

With the combination of the determined unit-cell volume and previously published v_P and v_S values at *P*–*T*s [23], the K_T, K_S, and *G* of diopside can be calculated via Equations (6)–(8); the results are shown in Table 2 and Figures 2 and 3.

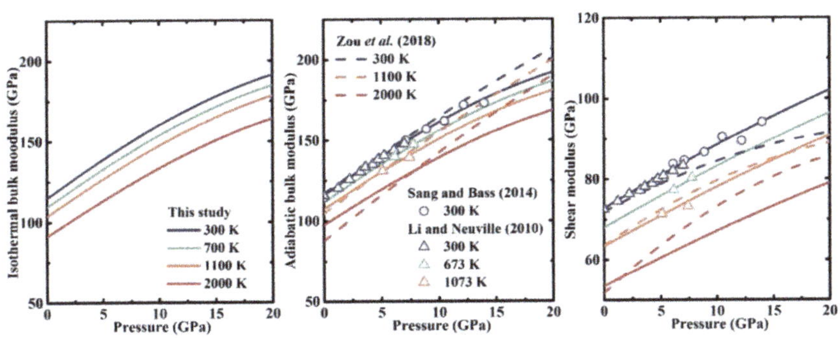

(a) (b) (c)

Figure 2. Elastic moduli of diopside as a function of pressure at various temperatures. Solid lines refer to our calculated results and dashed lines refer to the first-principle calculation results from Zou et al. (2018) [24]. The circles and triangles represent the results from Sang and Bass (2014) [37] and Li and Neuville (2010) [23], respectively. (**a**) Isothermal bulk modulus (**b**) Adiabatic bulk modulus (**c**) Shear modulus.

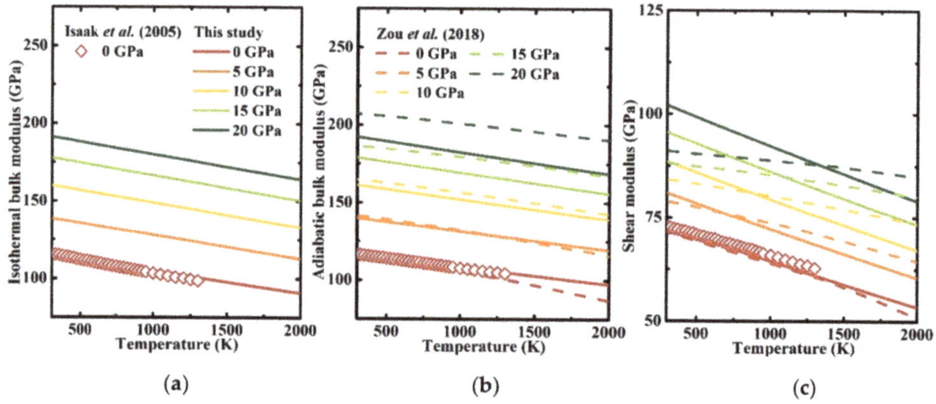

Figure 3. Elastic moduli of diopside as a function of temperature at various pressures. Solid lines refer to our calculated results and dashed lines refer to the first-principle calculation results from Zou et al. (2018) [24]. The diamonds correspond to the results from Isaak et al. (2006) [33]. (**a**) Isothermal bulk modulus (**b**) Adiabatic bulk modulus (**c**) Shear modulus.

As shown in Table 2, the K_T, K_S, and G values under ambient conditions have been determined as 114.8 GPa, 116 GPa, and 72.8 GPa, respectively, which generally agree with the experimental [23,25,37,42–45] and theoretical [24,46] results. All three elastic moduli show a nonlinear relationship with pressure and a linear relationship with temperature. At room temperature, the first pressure derivatives of the elastic moduli have been determined as $\partial K_T/\partial P = 5.17$, $\partial K_S/\partial P = 5.18$, and $\partial G/\partial P = 1.67$, and the second pressure derivatives of elastic moduli have been determined as $\partial^2 K_T/\partial P^2 = -0.066$/GPa, $\partial^2 K_S/\partial P^2 = -0.067$/GPa, and $\partial^2 G/\partial P^2 = -0.0106$/GPa. Also, the temperature derivatives of elastic moduli have been determined as $\partial K_T/\partial T = -0.0142$ GPa/K, $\partial K_S/\partial T = -0.011$ GPa/K, and $\partial G/\partial T = -0.0114$ GPa/K under ambient pressure conditions; these values are consistent with previously presented results [23–25].

Table 2. Elastic moduli of diopside and their first and second derivatives with respect to pressure and temperature.

K_{T0} GPa	$(\partial K_T/\partial P)_T$	$(\partial^2 K_T/\partial P^2)_T$ /GPa	$(\partial K_T/\partial T)_P$ 10^{-2} GPa/K	References
114 (4)	4.5 (1.8)			Levien and Prewitt (1981) [44]
109.1	4.84		−2.05	Zhao et al. (1998) [25]
105.1 (9)	6.8 (1)		-	Tribaudino et al. (2000) [43]
118 (1)	3.8 (2)			Thompson and Downs (2008) [42]
122	4.7			Walker et al. (2008) [46]
116.4	4.62		−1.72	Zou et al. (2018) [24]
114.8	5.17	−0.066	−1.42	This study
K_{S0} GPa	**$(\partial K_S/\partial P)_T$**	**$(\partial^2 K_S/\partial P^2)_T$ /GPa**	**$(\partial K_S/\partial T)_P$ 10^{-2} GPa/K**	**References**
116.4 (7)	4.9 (1)		−1.2 (1)	Li and Neuville (2010) [23]
114.6 (7)				Sang et al. (2011) [45]
114.6 (7)	4.8 (2)			Sang and Bass (2014) [37]
117.5	5.0	−0.026	−1.5	Zou et al. (2018)
116.0	5.18	−0.067	−1.10	This study
G_0 GPa	**$(\partial G/\partial P)_T$**	**$(\partial^2 G/\partial P^2)_T$ /GPa**	**$(\partial G/\partial T)_P$ 10^{-2} GPa/K**	**References**
73.0 (4)	1.6 (1)		−1.1 (1)	Li and Neuville (2010) [23]
72.7 (4)				Sang et al. (2011) [45]
72.7 (4)	1.7 (1)			Sang and Bass (2014) [37]
71.8	1.56	−0.0302	−0.871	Zou et al. (2018) [24]
72.8	1.67	−0.0106	−1.14	This study

The pressure and temperature dependences of these elastic moduli are shown in Figures 2 and 3, respectively. In Figure 2b,c, the calculated K_S and G values are in good agreement with the available experimental measurements under the relevant conditions [23,37]. Also, in Figure 2b, our ambient and high temperature K_S values generally agree with the first-principle calculations performed by Zou et al. [24] for under ~10 GPa. Then, the pressure derivative becomes smaller than that from Zou et al. [24], leading to a significant separation of ~10% at 20 GPa. Then, in Figure 2c, our calculated G value is in good agreement with the result from Zou et al. [24] at 1100 K. Compared to Zou et al. [24], our result is larger than that from Zou et al. [24] at 300 K, and smaller than that from Zou et al. [24] at 2000 K.

In Figure 3, at ambient pressure, our elastic moduli are generally consistent with the experimental study by Isaak et al. [33], which presented the ambient pressure elastic properties of chrome-diopside. The largest differences between our and Isaak's results are 2.1%, 0.6%, and 2.8% for K_T, K_S, and G, respectively. In addition, our K_S and G values are generally consistent with the results from Zou et al. [24] below 10 GPa. Then, the separations get more significant along with the pressure due to the differences in the pressure derivatives of K_S and G.

3.3. Thermodynamic Properties at High Temperatures and High Pressures

Our calculated thermal expansion, heat capacity, and Grüneisen parameters of diopside are illustrated in Figures 4–6, respectively. Figure 4a shows the temperature dependence of the thermal expansion of diopside at several pressure conditions, as well as previous results for comparisons. At ambient pressure, the temperature derivative of our calculated thermal expansion $\partial\alpha/\partial T$ is ~5.0 × 10^{-4}/K^2 over 500 K. This value is lower than that presented by Isaak et al. [33], which was obtained from the combination of previous studies [38,39,47]. Moreover, the difference in $\partial\alpha/\partial T$ causes a 17% deviation at 1300 K. On the other hand, our result is in good agreement with the recent data obtained via an X-ray diffraction study by Hovis et al. [29], with the largest difference being 4.7% in the temperature range of 500–1500 K.

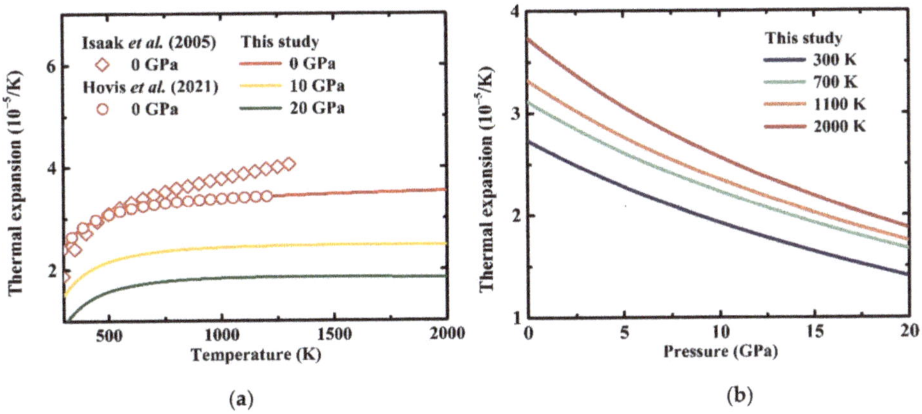

Figure 4. Temperature (**a**) and pressure (**b**) dependences of the thermal expansion of diopside. Solid lines refer to our calculated results. The diamonds and circles correspond to the results from Isaak et al. (2006) [33] and Hovis et al. (2021) [29], respectively.

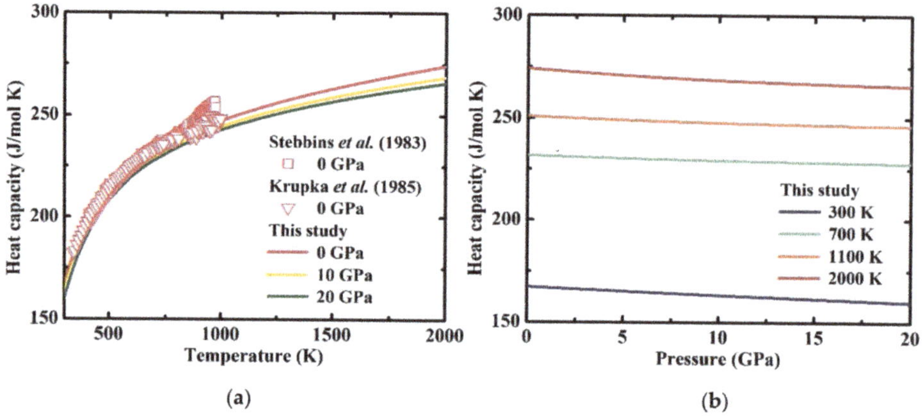

Figure 5. Temperature (**a**) and pressure (**b**) dependences of the heat capacity of diopside. Solid lines refer to our calculated results. The squares and triangles correspond to the results from Stebbins et al. (1983) [32] and Krupka et al. (1985) [30], respectively.

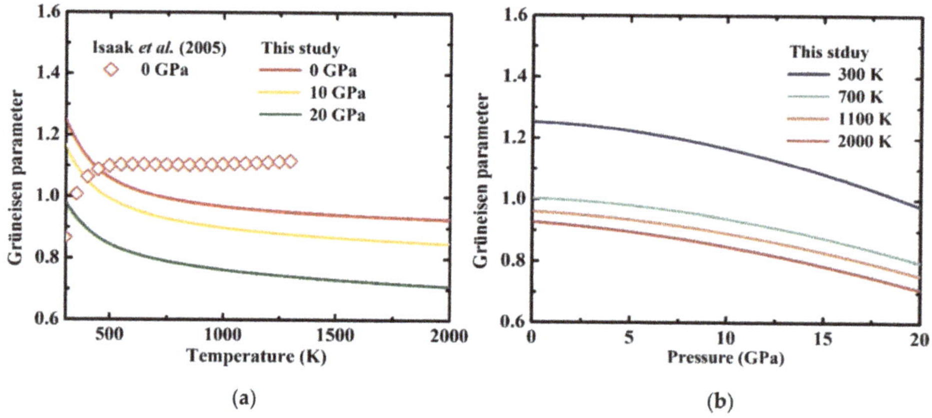

Figure 6. Temperature (**a**) and pressure (**b**) dependences of the Grüneisen parameter of diopside. Solid lines refer to our calculated results. The diamonds correspond to the results from Isaak et al. (2006) [33].

Currently, the published heat capacities of diopside are limited under ambient conditions. In Figure 5a, since the heat capacity of diopside for our calculation was taken from previous experimental studies [31], our obtained heat capacity is in good agreement with the experimental data presented by Krupka et al. [30], with a difference of 1.9%, and generally agrees with the result presented by Stebbins et al. [32] below ~850 K, with a difference of 1.5%.

The Grüneisen parameter is a valuable thermodynamic parameter in geophysics which can set limitations on the pressure and temperature dependence of the thermal properties of the mantle and core and constrain the adiabatic temperature gradient [48]. In Figure 6a, our calculated Grüneisen parameter shows a rapid decrease to ~500 K and then tends to nearly a constant, with a very small temperature derivative of -5.0×10^{-5}/K at 2000 K. The result is 15% lower than that from Isaak et al. [33] at 1300 K, which is probably caused by the difference in the chemical composition of minerals. Also, the Grüneisen parameter proposed by Isaak et al. [33] demonstrated a rapid increase below ~500 K; in contrast, our determined data shows an opposite trend, exhibiting a similar pattern to garnet and olivine under ambient pressure conditions [20,49].

Additionally, to investigate the effects of pressure on the thermodynamic properties of diopside, the calculated thermal expansion, heat capacity, and Grüneisen parameters as a function of pressure at different temperatures are illustrated in Figures 4b, 5b and 6b, respectively. All α, C_P, and γ values for diopside show nonlinear pressure dependences. Hence, we fitted the thermodynamic parameters into the equation $N = N_0 + \partial N/\partial P \times P + \partial^2 N/\partial P^2 \times P^2$ at different temperatures; where N refers to the thermodynamic properties; N_0 refers to N ambient pressure; and $\partial N/\partial P$ and $\partial^2 N/\partial P^2$ refer to the first and second pressure derivatives of N, respectively. The fitting coefficients are listed in Table 3.

Table 3. The thermal expansion, heat capacity, and Grüneisen parameters of diopside and their first and second derivatives with respect to pressure.

T K	α_0 10^{-5}/K	$\partial\alpha/\partial P$ 10^{-6}/K GPa	$\partial^2\alpha/\partial P^2$ 10^{-8}/K GPa^2
300	2.71	−0.92	1.4
700	3.09	−1.02	1.6
1100	3.29	−1.13	1.8
2000	3.70	−1.37	2.4
T K	**C_{P0} J/mol K**	**$\partial C_P/\partial P$ J/mol K GPa**	**$\partial^2 C_P/\partial P^2$ 10^{-3} J/mol K GPa^2**
300	167.15	−0.417	2.3
700	231.57	−0.275	4.5
1100	250.91	−0.372	7.2
2000	274.09	−0.688	14.2
T K	**γ_0**	**$\partial\gamma/\partial P$ 10^{-3}/GPa**	**$\partial^2\gamma/\partial P^2$ 10^{-4}/GPa^2**
300	1.254	−3.4	−5.2
700	1.004	−2.7	−3.8
1100	0.961	−3.4	−3.5
2000	0.928	−4.8	−3.1

In Table 3, all α, C_P, and γ values show negative pressure dependences, and the effects of pressure on α and C_P increase along with the temperature. At room temperature, the first pressure derivative of α is $\partial\alpha/\partial P = -0.92 \times 10^{-6}$/K GPa. This value is larger than that suggested by Zhao et al. [25] ($-1.52(2) \times 10^{-6}$/K GPa), which was determined in the P–T range of 8.2 GPa and 1280 K with a linear fitting equation. Then, increasing the temperature to 2000 K, $\partial\alpha/\partial P$ decreases to $\partial\alpha/\partial P = -1.37 \times 10^{-6}$/K GPa. Similarly, the pressure derivative of C_P ($\partial C_P/\partial P$) decreases from −0.417 J/mol K GPa to −0.688 J/mol K GPa in the temperature range of 300–2000 K. In particular, the first pressure derivative of the Grüneisen parameter is $\partial\gamma/\partial P = -3.4 \times 10^{-3}$/GPa at room temperature. Then, the value increases with temperature to -2.6×10^{-3}/GPa at ~500 K because of the variations at the low-temperature range, and decreases with the temperature again to -4.8×10^{-3}/GPa at 2000 K.

4. Discussion

4.1. *Effect of Mineral Composition on the Adiabatic Temperature Gradient*

Based on the obtained thermodynamic data on diopside, we have estimated the adiabatic temperature gradient ($(\partial T/\partial z)_S$) by using Equation (1); the result is shown in Figure 7. To investigate the effect of mineral composition on the adiabatic temperature gradient, the $(\partial T/\partial z)_S$ values obtained using thermodynamic data on Fe-bearing olivine ($[Fe_{0.2}Mg_{0.8}]_2SiO_4$), pyrope ($Mg_3Al_2[SiO_4]_3$), almandine ($Fe_3Al_2[SiO_4]_3$), and grossular ($Ca_3Al_2[SiO_4]_3$) are also illustrated in Figure 7 for comparison [19–21]. Note that the mantle potential temperature (T_P), which refers to the temperature that it would have upon ascending adiabatically to the Earth's surface without undergoing partial melting [50], is employed as T_P = 1610 K [18]. Additionally, the pressure and gravitational acceleration variations with depth are from the study performed by Stacey and Davis [4].

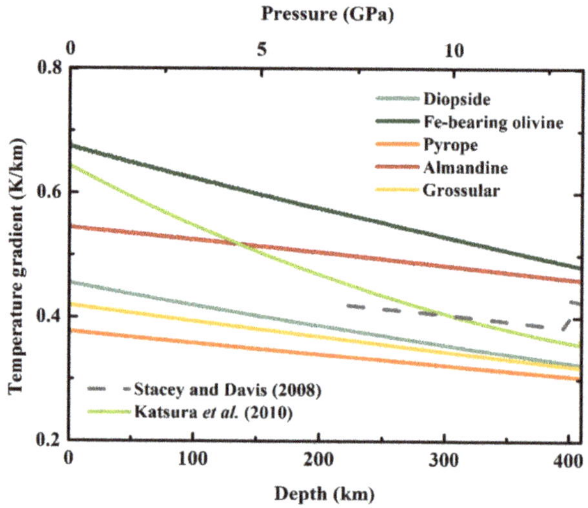

Figure 7. Adiabatic temperature gradients obtained using thermodynamic data on minerals (T_P = 1610 K).

The $(\partial T/\partial z)_S$ values obtained using thermodynamic data on diopside are 0.46 K/km, 0.39 K/km, and 0.32 K/km at depths of 0 km, 200 km, and 410 km, respectively; the value is ~14% lower than that obtained by Stacey and Davis [4] in the depth range of 220–410 km. For the Fe-free minerals, at a depth of 200 km, the $(\partial T/\partial z)_S$ value obtained using thermodynamic data on diopside is 14% and 5% higher than that of pyrope and grossular, respectively, whilst it is 17% lower than that of forsterite. Also, this value is 33% and 23% lower than that of the Fe-bearing olivine and Fe-bearing garnet (almandine), respectively. Moreover, a previous study suggested that the depth derivatives of $(\partial T/\partial z)_S$ of those minerals from the same group are similar [20]. In this study, the depth derivative of $(\partial T/\partial z)_S$ obtained from thermodynamic data on diopside is -3.0×10^{-4} K/km^2 in the depth range of 200–410 km, which is smoother than that of olivine (-5.0×10^{-4} K/km^2) and steeper than that of garnet (-2.0×10^{-4} K/km^2) [20].

4.2. Adiabatic Geotherm of the Eclogite Model

It is widely accepted that the bulk mantle mineral assemblages can be represented by the pyrolite model [11,51]. The seismic discontinuities at depths of 410 km, 520 km, and 660 km are associated with the phase transitions from $(Mg,Fe)_2SiO_4$ olivine (α-phase) to wadsleyite (β-phase) and from wadsleyite to ringwoodite (γ-phase) and the dissociation of ringwoodite into bridgmanite and periclase, respectively, for the pyrolite model [52,53]. Hence, the temperature at these seismic discontinuities can be estimated by comparing the depth of these seismic discontinuities with the transition pressures in the system Mg_2SiO_4-Fe_2SiO_4 [2,18]. With the combination of the thermal equations of states of $(Mg,Fe)_2SiO_4$ olivine and its high-pressure phase minerals, as well as Equation (1), the adiabatic temperature gradient $(\partial T/\partial z)_S$ and adiabatic geotherm of a pyrolitic mantle was presented for ~3000 km by Katsura et al. [18], which has been widely used in later investigations.

Meanwhile, the upper mantle under the subducted oceanic crust is generally considered eclogitic in composition [12,54]. The subducting slab has a layered structure: overlying crust of mid-ocean ridge basalt (MORB), followed by residual harzburgite and depleted pyrolite from top to bottom [55,56]. When the subducting slab is subducted into the mantle, the MORB transforms directly into eclogite, which mainly contains an assemblage of eclogitic garnet and clinopyroxene, and becomes denser than the harzburgite underlying it [57,58]. Then, the clinopyroxene starts to dissolve progressively into stishovite and garnet, which then completely forms majoritic garnet by a depth of ~450 km [59–61]. Hence, the mineral composition of an eclogitic upper mantle is quite different from a pyrolitic upper mantle, which raises the question: Will the geotherm of the eclogite model also be different from the pyrolite model?

In order to estimate the adiabatic geotherm of an eclogitic upper mantle with Equation (1), it is necessary to determine the T_P of the oceanic mantle first. Previous research on the generation of the oceanic upper mantle proposed the T_P of mid-ocean ridges as 1533 K [62]; then, the temperature was suggested to be hotter [63,64]. Recently, with the consideration of water content of the peridotite in MORB, the T_P of mid-ocean ridges was constrained to 1623–1683 K [65]. Also, the composition at the top of an eclogitic upper mantle (200 km) is ~70% clinopyroxene, and ~30% eclogitic garnet [66]. Therefore, with the high temperature and high pressure thermodynamic data on diopside and eclogitic garnet, the temperature at a depth of 200 km can be estimated as 1699–1762 K via Equation (13) [50]. It is worth noting that the eclogitic garnet is composed of approximately 40% pyrope, 30% almandine, and 30% grossular [67,68].

$$T_P = Texp\left(-\frac{g\alpha z}{C_P}\right) \tag{13}$$

Based on the variations of mineral volume fraction with depth presented by Irifune et al. [66], our calculated $(\partial T/\partial z)_S$ value of the eclogite model in the depth range of 200–450 km is shown in Figure 8, together with the result obtained by using the seismic velocity [4] and the result for a pyrolitic upper mantle [18]. The $(\partial T/\partial z)_S$ value of the upper mantle was presented as a constant of 0.3 K/km in an early study [5]; later, a higher value was suggested at the top part of the upper mantle (0.4–0.5 K/km) [4,18]. In Figure 8, the $(\partial T/\partial z)_S$ value of the eclogitic model decreases from 0.39 K/km to 0.34 K/km at a depth range of 200–410 km when T_P = 1623 K, which is ~12% lower than that presented by Stacey and Davis [4]. Also, the $(\partial T/\partial z)_S$ value of the eclogite model is ~17% lower than that of the pyrolite model suggested by Katsura et al. [18] at 200 km, but increasing the depth to 410 km, the values become closer, with a difference of 4%. Under 410 km, the $(\partial T/\partial z)_S$ value of the eclogite model maintains the same trend as the depth and decreases to 0.33 K/km at 450 km, while the $(\partial T/\partial z)_S$ value of the pyrolite model shows an increase because of the olivine–wadsleyite transition. Moreover, the value of $(\partial T/\partial z)_S$ for the eclogite model shows a ~5% increase with T_P increasing from 1623 K to 1683 K; the values are 0.41 K/km, 0.35 K/km, and 0.34 K/km at depths of 200 km, 410 km, and 450 km, respectively.

Our obtained adiabatic geotherm of an eclogitic mantle from 200 km to 450 km is shown in Figure 9. At depths of 200 km, 410 km, and 450 km, the temperatures of the eclogite model are 1699 K, 1775 K, and 1789 K when T_P = 1623 K, respectively, and 1762 K, 1841 K, and 1855 K when T_P = 1683 K, respectively. Under the condition of T_P = 1683 K, the estimated temperature is generally consistent with the results proposed by Stacey and Davis [4] and Trubitsyn and Trubitsyn [69] in the depth range of ~200–410 km, with the differences being less than 0.8% and 1%, respectively, but ~80 K higher than the results obtained by Anderson [8] and Brown and Shankland [9] at a depth of 410 km. By comparing the depth of the 410 km discontinuity with the forsterite–wadsleyite phase transition pressure, the temperature at a depth of 409 km was determined to be 1830 $\pm$ 48 K for the pyrolite model, and consequently, the T_P was presented as 1610 $\pm$ 35 K [18]. Because of the existence of olivine in the pyrolitic mantle, though the T_P of the pyrolite model is lower than that of the eclogite model, the higher $(\partial T/\partial z)_S$ value of the pyrolite model leads to similar temperatures to the eclogite model in the depth range of 200–410 km. However, the temperature of the eclogite model does not show the increase demonstrated in the pyrolite model at the 410 km discontinuity, but maintains the same trend at a depth of 450 km.

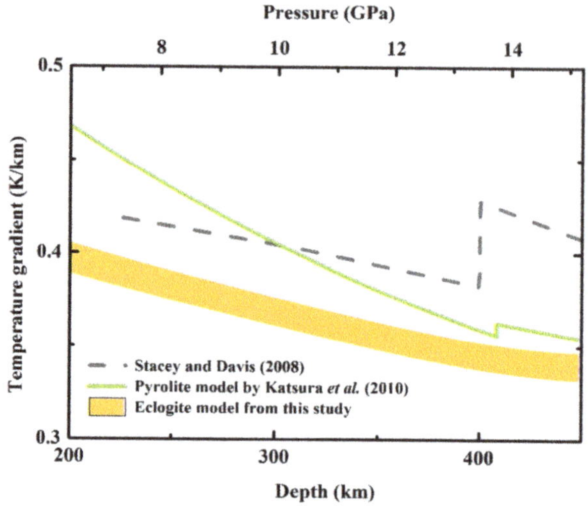

Figure 8. Adiabatic temperature gradients in the depth range of 200–450 km. The lower and upper edges of the filled area are calculated with T_P as 1623 K and 1683 K, respectively.

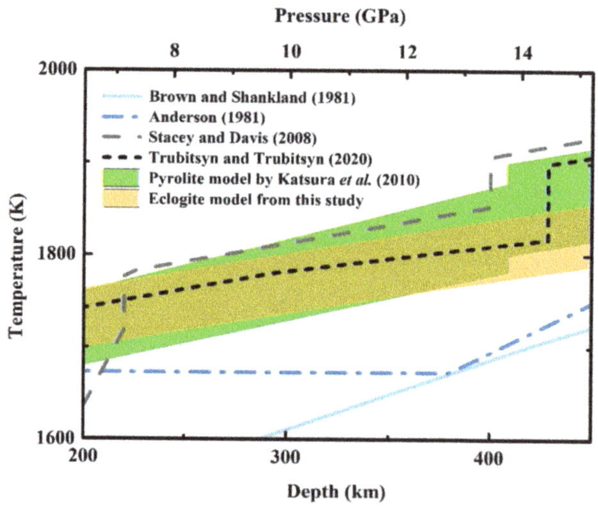

Figure 9. Adiabatic geotherm in the depth range of 200–450 km. The lower and upper edges of the filled area are calculated with T_P as 1623 K and 1683 K, respectively.

It is worth noting that, except for olivine, the peridotitic garnet, clinopyroxene, and orthopyroxene are essential compositions in a pyrolite mantle [67,70]. Additionally, the majority of the clinopyroxene in an eclogite mantle is considered to be omphacite, which is mainly composed of diopside and jadeite ($NaAlSi_2O_6$) [71]. Also, the majoritic garnet becomes dominant over a depth of ~450 km [59–61]. Furthermore, Fe is a non-negligible content in both olivine and clinopyroxene [72,73]. Previous studies have suggested that Fe incorporation could increase the estimated adiabatic temperature gradient [20]. Hence, the differences between the geotherm of the pyrolite and eclogite models are probably not the same as what we described if we consider more complicated mineral compositions for the pyrolite model. More thermodynamic parameters of mantle minerals are still required to improve the geotherm model. Overall, this study shows that the adiabatic temperature gradient and adiabatic geotherm of different mantle compositions exhibit various trends with depth. As a consequence, suitable geotherms need to be taken into account when investigating different tectonic settings.

5. Conclusions

In summary, following a numerical iterative procedure, we obtained self-consistent thermodynamic parameters: thermal expansion, heat capacity, and Grüneisen parameters of diopside in a wide *P*–*T* range (20 GPa, 2000 K). All these thermodynamic parameters are nonlinearly and negatively correlated with pressure. Also, the pressure effects on thermal expansion and heat capacity increase with temperature, while the effect of pressure on the Grüneisen parameter decreases to a minimum at ~500 K, then increases with temperature again. Based on thermodynamic data on major mantle minerals from this and previous studies, we have estimated the adiabatic temperature gradient and geotherm of an eclogitic upper mantle. The results suggest that the adiabatic temperature gradient of the eclogite model is lower than that of the pyrolite model at the same depth, which causes a slower rise of temperature than that of a pyrolitic upper mantle and leads to a similar adiabatic geotherm for both pyrolitic and eclogitic upper mantle in the depth range of 200–410 km.

Supplementary Materials: The following are available online at https://www.mdpi.com/article/10.3390/min11121322/s1, File S1: MS Excel spreadsheet for unit-cell volume of diopside (unit-cell volume.xlsx), File S2: MS Excel spreadsheet for isothermal bulk modulus of diopside (isothermal bulk modulus.xlsx), File S3: MS Excel spreadsheet for adiabatic bulk modulus of diopside (adiabatic bulk modulus.xlsx), File S4: MS Excel spreadsheet for shear modulus of diopside (shear modulus.xlsx), File S5: MS Excel spreadsheet for thermal expansion of diopside (thermal expansion.xlsx), File S6: MS Excel spreadsheet for heat capacity of diopside (heat capacity.xlsx), File S7: MS Excel spreadsheet for Grüneisen parameter of diopside (Grüneisen parameter.xlsx).

Author Contributions: Conceptualization, C.S., Y.L. and D.F.; methodology, C.S., Y.L. and D.F.; software, G.Y.; resources, J.J., Z.S., Y.W. and W.Q.; data curation, C.S.; writing—original draft preparation, C.S.; writing—review and editing, C.S., Y.L., D.F. and W.S.; supervision, Y.L. and D.F.; project administration, Y.L. and D.F.; funding acquisition, C.S. and Y.L. All authors have read and agreed to the published version of the manuscript.

Funding: This research was funded by the Science and Technology Project for the Langfang City [2021013079] and the National Natural Science Foundation of China [41873075].

Data Availability Statement: Data is contained within the article or supplementary material.

Acknowledgments: The authors thank Wenge Zhou at the Institute of Geochemistry, Chinese Academy of Sciences for helpful discussions and suggestions. We also thank Peter I. Dorogokupets at the Institute of the Earth Crust SB RAS and two anonymous reviewers for their insightful comments and suggestions that were very valuable for improving our manuscript.

Conflicts of Interest: The authors declare no conflict of interest.

References

1. Weidner, D.J.; Wang, Y. Phase transformations: Implications for mantle structure. In *Earth's Deep Interior: Mineral. Physics and Tomography From the Atomic to the Global Scale*; American Geophysical Union: Washington, DC, USA, 2000; pp. 215–235.
2. Akaogi, M.; Ito, E.; Navrotsky, A. Olivine-modified spinel-spinel transitions in the system Mg_2SiO_4-Fe_2SiO_4: Calorimetric measurements, thermochemical calculation, and geophysical application. *J. Geophys. Res. Solid Earth* **1989**, *94*, 15671–15685. [CrossRef]
3. Ono, S. Experimental constraints on the temperature profile in the lower mantle. *Phys. Earth Planet. Inter.* **2008**, *170*, 267–273. [CrossRef]
4. Stacey, F.; Davis, P. *Physics of the Earth*; Cambridge Univ. Press: Cambridge, UK, 2008.
5. Turcotte, D.L.; Schubert, G. *Geodynamics*, 2nd ed.; Cambridge University Press: Cambridge, UK, 2012.
6. Dziewonski, A.M.; Hales, A.L.; Lapwood, E.R. Parametrically simple earth models consistent with geophysical data. *Phys. Earth Planet. Inter.* **1975**, *10*, 12–48. [CrossRef]
7. Stacey, F.D. A thermal model of the earth. *Phys. Earth Planet. Inter.* **1977**, *15*, 341–348. [CrossRef]
8. Anderson, O.L. Temperature profiles in the Earth. *Evol. Earth* **1981**, *5*, 19–27.

9. Brown, J.M.; Shankland, T.J. Thermodynamic parameters in the Earth as determined from seismic profiles. *Geophys. J. Int.* **1981**, *66*, 579–596. [CrossRef]
10. Jeanloz, R.; Richter, F.M. Convection, composition, and the thermal state of the lower mantle. *J. Geophys. Res.* **1979**, *84*, 5497–5504. [CrossRef]
11. Ringwood, A.E. A model for the upper mantle. *J. Geophys. Res.* **1962**, *67*, 4473–4478. [CrossRef]
12. Bass, J.D.; Anderson, D.L. Composition of the upper mantle: Geophysical tests of two petrological models. *Geophys. Res. Lett.* **1984**, *11*, 229–232. [CrossRef]
13. Katsura, T. Thermal expansion of Mg_2SiO_4 ringwoodite at high pressures. *J. Geophys. Res.* **2004**, *109*, 209–218. [CrossRef]
14. Katsura, T.; Shatskiy, A.; Manthilake, M.A.G.M.; Zhai, S.; Fukui, H.; Yamazaki, D.; Matsuzaki, T.; Yoneda, A.; Ito, E.; Kuwata, A.; et al. Thermal expansion of forsterite at high pressures determined by in situ X-ray diffraction: The adiabatic geotherm in the upper mantle. *Phys. Earth Planet. Inter.* **2009**, *174*, 86–92. [CrossRef]
15. Katsura, T.; Shatskiy, A.; Manthilake, M.A.G.M.; Zhai, S.; Yamazaki, D.; Matsuzaki, T.; Yoshino, T.; Yoneda, A.; Ito, E.; Sugita, M.; et al. P-V-T relations of wadsleyite determined by in situ X-ray diffraction in a large-volume high-pressure apparatus. *Geophys. Res. Lett.* **2009**, *36*, 1–5.
16. Katsura, T.; Yamada, H.; Nishikawa, O.; Song, M.; Kubo, A.; Shinmei, T.; Yokoshi, S.; Aizawa, Y.; Yoshino, T.; Walter, M.J.; et al. Olivine-wadsleyite transition in the system $(Mg,Fe)_2SiO_4$. *J. Geophys. Res. Solid Earth* **2004**, *109*, 1–12. [CrossRef]
17. Katsura, T.; Yokoshi, S.; Kawabe, K.; Shatskiy, A.; Manthilake, M.A.G.M.; Zhai, S.; Fukui, H.; Hegoda, H.A.C.I.; Yoshino, T.; Yamazaki, D.; et al. P-V-T relations of $MgSiO_3$ perovskite determined by in situ X-ray diffraction using a large-volume high-pressure apparatus. *Geophys. Res. Lett.* **2009**, *36*, 1–6.
18. Katsura, T.; Yoneda, A.; Yamazaki, D.; Yoshino, T.; Ito, E. Adiabatic temperature profile in the mantle. *Phys. Earth Planet. Inter.* **2010**, *183*, 212–218. [CrossRef]
19. Su, C.; Liu, Y. Thermal expansion, heat capacity and Grüneisen parameter of grossular at high temperature and high pressure. *High. Temp.-High. Press.* **2021**, *50*, 105–119.
20. Su, C.; Liu, Y.; Fan, D.; Song, W.; Yang, G. Self-consistent thermodynamic parameters of pyrope and almandine at high-temperature and high-pressure conditions: Implication on the adiabatic temperature gradient. *Phys. Earth Planet. Inter.* **2021**, 106789. [CrossRef]
21. Su, C.; Liu, Y.; Song, W.; Fan, D.; Wang, Z.; Tang, H. Thermodynamic properties of San Carlos olivine at high temperature and high pressure. *Acta Geochim.* **2018**, *37*, 171–179. [CrossRef]
22. Ringwood, A.E. Phase transformations and the constitution of the mantle. *Phys. Earth Planet. Inter.* **1970**, *3*, 109–155. [CrossRef]
23. Li, B.; Neuville, D.R. Elasticity of diopside to 8 GPa and 1073 K and implications for the upper mantle. *Phys. Earth Planet. Inter.* **2010**, *183*, 398–403. [CrossRef]
24. Zou, F.; Wu, Z.; Wang, W.; Wentzcovitch, R.M. An extended semianalytical approach for thermoelasticity of monoclinic crystals: Application to diopside. *J. Geophys. Res. Solid Earth* **2018**, *123*, 7629–7643. [CrossRef]
25. Zhao, Y.; Dreele, R.B.V.; Zhang, J.Z.; Weidner, D.J. Thermoelastic Equation of State of Monoclinic Pyroxene: $CaMgSi_2O_6$ Diopside. *Rev. High. Press. Sci. Technol.* **1998**, *7*, 25–27. [CrossRef]
26. Birch, F. Finite Elastic Strain of Cubic Crystals. *Phys. Rev.* **1947**, *71*, 809–824. [CrossRef]
27. Jackson, I.; Rigden, S.M. Analysis of P-V-T data: Constraints on the thermoelastic properties of high-pressure minerals. *Phys. Earth Planet. Inter.* **1996**, *96*, 85–112. [CrossRef]
28. Pandolfo, F.; Cámara, F.; Domeneghetti, M.C.; Alvaro, M.; Nestola, F.; Karato, S.-I.; Amulele, G. Volume thermal expansion along the jadeite–diopside join. *Phys. Chem. Miner.* **2015**, *42*, 1–14. [CrossRef]
29. Hovis, G.L.; Tribaudino, M.; Leaman, A.; Almer, C.; Altomare, C.; Morris, M.; Maksymiw, N.; Morris, D.; Jackson, K.; Scott, B.; et al. Thermal expansion of minerals in the pyroxene system and examination of various thermal expansion models. *Am. Mineral.* **2021**, *106*, 883–899. [CrossRef]
30. Krupka, K.M.; Hemingway, B.S.; Robie, R.A.; Kerrick, D.M. High-temperature heat capacities and derived thermodynamic properties of anthophyllite, diopside, dolomite, enstatite, bronzite, talc, tremolite and wollastonite. *Am. Mineral.* **1985**, *70*, 261–271.
31. Richet, P.; Fiquet, G. High-temperature heat capacity and premelting of minerals in the system $MgO\text{-}CaO\text{-}Al_2O_3\text{-}SiO_2$. *J. Geophys. Res.* **1991**, *96*, 445–456. [CrossRef]
32. Stebbins, J.F.; Carmichael, I.S.E.; Weill, D.E. The high temperature liquid and glass heat contents and the heats of fusion of diopside, albite, sanidine and nepheline. *Am. Mineral.* **1983**, *68*, 717–730.
33. Isaak, D.G.; Ohno, I.; Lee, P.C. The elastic constants of monoclinic single-crystal chrome-diopside to 1300 K. *Phys. Chem. Miner.* **2006**, *32*, 691–699. [CrossRef]
34. De Koker, N.; Stixrude, L. Self-consistent thermodynamic description of silicate liquids, with application to shock melting of MgO periclase and $MgSiO_3$ perovskite. *Geophys. J. Int.* **2009**, *178*, 162–179. [CrossRef]
35. Sokolova, T.S.; Dorogokupets, P.I. Equations of State of Ca-Silicates and Phase Diagram of the $CaSiO_3$ System under Upper Mantle Conditions. *Minerals* **2021**, *11*, 322. [CrossRef]
36. Angel, R.J. Equations of State. *Rev. Mineral. Geochem.* **2000**, *41*, 35–59. [CrossRef]
37. Sang, L.; Bass, J.D. Single-crystal elasticity of diopside to 14 GPa by Brillouin scattering. *Phys. Earth Planet. Inter.* **2014**, *228*, 75–79. [CrossRef]

38. Cameron, M.; Sueno, S.; Prewitt, C.T.; Papike, J.J. High-Temperature Crystal Chemistry of Acmite, Diopside, Hedenbergite Jadeite, Spodumene and Ureyite. *Am. Mineral.* **1973**, *58*, 594–618.
39. Finger, L.W.; Ohashi, Y. The thermal expansion of diopside to 800 °C and a refinement of the crystal structure at 700 °C. *Am. Mineral.* **1976**, *61*, 303–310.
40. Gonzalez-Platas, J.; Alvaro, M.; Nestola, F.; Angel, R. EosFit7-GUI: A new graphical user interface for equation of state calculations, analyses and teaching. *J. Appl. Crystallogr.* **2016**, *49*, 1377–1382. [CrossRef]
41. Fei, Y. Thermal Expansion. In *Mineral. Physics & Crystallography*; Ahrens, T.J., Ed.; American Geophysical Union: Washington, DC, USA, 1995; pp. 29–44.
42. Thompson, R.M.; Downs, R.T. The crystal structure of diopside at pressure to 10 GPa. *Am. Mineral.* **2008**, *93*, 177–186. [CrossRef]
43. Tribaudino, M.; Prencipe, M.; Bruno, M.; Levy, D. High-pressure behaviour of Ca-rich *C2/c* clinopyroxenes along the join diopside-enstatite ($CaMgSi_2O_6$-$Mg_2Si_2O_6$). *Phys. Chem. Miner.* **2000**, *27*, 656–664. [CrossRef]
44. Levien, L.; Prewitt, C.T. High-pressure structural study of diopside. *Am. Mineral.* **1981**, *66*, 315–323.
45. Sang, L.; Vanpeteghem, C.B.; Sinogeikin, S.V.; Bass, J.D. The elastic properties of diopside, $CaMgSi_2O_6$. *Am. Mineral.* **2011**, *96*, 224–227. [CrossRef]
46. Walker, A.M.; Tyer, R.P.; Bruin, R.P.; Dove, M.T. The compressibility and high pressure structure of diopside from first principles simulation. *Phys. Chem. Miner.* **2008**, *35*, 359–366. [CrossRef]
47. Saxena, S.K.; Chatterjee, N.; Fei, Y.; Shen, G. *Thermodynamic Data on Oxides and Silicates. An Assessed Data Set Based on Thermochemistry and High Pressure Phase Equilibrium*, 1st ed.; Springer: Berlin/Heidelberg, Germany, 1993.
48. Vočadlo, N.L.; Price, G.D. The Grüneisen parameter-computer calculations via lattice dynamics. *Phys. Earth Planet. Inter.* **1994**, *82*, 261–270. [CrossRef]
49. Wu, Z.; Wentzcovitch, R.M. Vibrational and thermodynamic properties of wadsleyite: A density functional study. *J. Geophys. Res.* **2007**, *112*. [CrossRef]
50. McKenzie, D.; Bickle, M.J. The volume and composition of melt generated by extension of the lithosphere. *J. Petrol.* **1988**, *29*, 625–679. [CrossRef]
51. Ringwood, A.E. Phase transformations and their bearing on the constitution and dynamics of the mantle. *Geochim. Et Cosmochim. Acta* **1991**, *55*, 2083–2110. [CrossRef]
52. Ringwood, A.E. Phase transformations in the mantle. *Earth Planet. Sci. Lett.* **1968**, *5*, 401–412. [CrossRef]
53. Muir, J.M.R.; Zhang, F.; Brodholt, J.P. The effect of water on the post-spinel transition and evidence for extreme water contents at the bottom of the transition zone. *Earth Planet. Sci. Lett.* **2021**, *565*, 116909. [CrossRef]
54. Ringwood, A.E. Phase Transformations and Differentiation in Subducted Lithosphere: Implications for Mantle Dynamics, Basalt Petrogenesis, and Crustal Evolution. *J. Geol.* **1982**, *90*, 611–643. [CrossRef]
55. Ringwood, A.E.; Irifune, T. Nature of the 650–km seismic discontinuity: Implications for mantle dynamics and differentiation. *Nature* **1988**, *331*, 131–136. [CrossRef]
56. Ishii, T.; Kojitani, H.; Akaogi, M. Phase Relations of Harzburgite and MORB up to the Uppermost Lower Mantle Conditions: Precise Comparison with Pyrolite by Multisample Cell High-Pressure Experiments with Implication to Dynamics of Subducted Slabs. *J. Geophys. Res. Solid Earth* **2019**, *124*, 3491–3507. [CrossRef]
57. Green, D.H.; Ringwood, A.E. An experimental investigation of the gabbro to eclogite transformation and its petrological applications. *Geochim. Et Cosmochim. Acta* **1967**, *31*, 767–833. [CrossRef]
58. Poli, S. The amphibolite-eclogite transformation; an experimental study on basalt. *Am. J. Sci.* **1993**, *293*, 1061–1107. [CrossRef]
59. Akaogi, M.; Akimoto, S. Pyroxene-garnet solid-solution equilibria in the systems $Mg_4Si_4O_{12}$-$Mg_3Al_2Si_3O_{12}$ and $Fe_4Si_4O_{12}$-$Fe_3Al_2Si_3O_{12}$ at high pressures and temperatures. *Phys. Earth Planet. Inter.* **1977**, *15*, 90–106. [CrossRef]
60. Irifune, T. Phase transformations in the earth's mantle and subducting slabs: Implications for their compositions, seismic velocity and density structures and dynamics. *Isl. Arc.* **1993**, *2*, 55–71. [CrossRef]
61. Kiseeva, E.S.; Yaxley, G.M.; Stepanov, A.S.; Tkalčić, H.; Litasov, K.D.; Kamenetsky, V.S. Metapyroxenite in the mantle transition zone revealed from majorite inclusions in diamonds. *Geology* **2013**, *41*, 883–886. [CrossRef]
62. Presnall, D.C.; Gudfinnsson, G.H.; Walter, M.J. Generation of mid-ocean ridge basalts at pressures from 1 to 7 GPa. *Geochim. Et Cosmochim. Acta* **2002**, *66*, 2073–2090. [CrossRef]
63. Putirka, K.D. Mantle potential temperatures at Hawaii, Iceland, and the mid-ocean ridge system, as inferred from olivine phenocrysts: Evidence for thermally driven mantle plumes. *Geochem. Geophys. Geosystems* **2005**, *6*, 1–14. [CrossRef]
64. Dalton, C.A.; Langmuir, C.H.; Gale, A. Geophysical and geochemical evidence for deep temperature variations beneath mid-ocean ridges. *Science* **2014**, *344*, 80–83. [CrossRef] [PubMed]
65. Sarafian, E.; Gaetani, G.A.; Hauri, E.H.; Sarafian, A.R. Experimental constraints on the damp peridotite solidus and oceanic mantle potential temperature. *Science* **2017**, *355*, 942–945. [CrossRef] [PubMed]
66. Irifune, T.; Sekine, T.; Ringwood, A.E.; Hibberson, W.O. The eclogite-garnetite transformation at high pressure and some geophysical implications. *Earth Planet. Sci. Lett.* **1986**, *77*, 245–256. [CrossRef]
67. Duan, L.; Wang, W.; Wu, Z.; Qian, W. Thermodynamic and Elastic Properties of Grossular at High Pressures and High Temperatures: A First-Principles Study. *J. Geophys. Res. Solid Earth* **2019**, *124*, 7792–7805. [CrossRef]

68. Xu, J.; Zhang, D.; Fan, D.; Dera, P.K.; Shi, F.; Zhou, W. Thermoelastic Properties of Eclogitic Garnets and Omphacites: Implications for Deep Subduction of Oceanic Crust and Density Anomalies in the Upper Mantle. *Geophys. Res. Lett.* **2019**, *46*, 179–188. [CrossRef]
69. Trubitsyn, A.P.; Trubitsyn, V.P. Temperature Distribution in the Earth's Mantle. *Dokl. Earth Sci.* **2020**, *495*, 905–909. [CrossRef]
70. Green, D.H.; Ringwood, A.E. Mineralogy of peridotitic compositions under upper mantle conditions. *Phys. Earth Planet. Inter.* **1970**, *3*, 359–371. [CrossRef]
71. Liu, L.-G. The mineralogy of an eclogitic earth mantle. *Phys. Earth Planet. Inter.* **1980**, *23*, 262–267. [CrossRef]
72. Ringwood, A.E.; Major, A. The system Mg_2SiO_4-Fe_2SiO_4 at high pressures and temperatures. *Phys. Earth Planet. Inter.* **1970**, *3*, 89–108. [CrossRef]
73. Canil, D.; O'Neill, H.S.C. Distribution of Ferric Iron in some Upper-Mantle Assemblages. *J. Petrol.* **1996**, *37*, 609–635. [CrossRef]

minerals

MDPI

Article

Equation of State, Compressibility, and Vibrational Properties of Brucite over Wide Pressure and Temperature Ranges: Atomistic Computer Simulations with the Modified ClayFF Classical Force Field

Evgeny V. Tararushkin [1], Vasily V. Pisarev [1,2] and Andrey G. Kalinichev [3,*]

1 International Laboratory for Supercomputer Atomistic Modelling and Multi-Scale Analysis, HSE University, 123458 Moscow, Russia
2 Joint Institute for High Temperatures of the Russian Academy of Sciences, 125412 Moscow, Russia
3 Laboratoire SUBATECH, UMR 6457–IMT Atlantique, Nantes Université, CNRS/IN2P3, 44307 Nantes, France
* Correspondence: kalinich@subatech.in2p3.fr

Abstract: The behavior of brucite over wide ranges of temperatures and pressures is of great interest for fundamental geochemistry and geophysics. Brucite layers and their octahedral $Mg(OH)_6$ structural units constitute an important structural part of layered dense magnesium hydrous silicates (DMHS), which play a major role in mineral equilibria controlling water balance in the subduction zones of the upper mantle. The ClayFF force field was originally developed for atomistic computer simulations of clays and other layered minerals and their hydrated interfaces. The crystallographic parameters of brucite at 25 °C and 1 bar were used, among several others, to develop the original ClayFF parametrization. Its new recent modification, ClayFF-MOH, can more accurately account for the bending of Mg–O–H angles in the brucite structure, and it was used here to test the applicability of this simple classical model over very wide ranges of temperature and pressure well beyond the range of its original implementation (up to 600 °C and 15 GPa). The pressure and temperature dependencies of brucite crystallographic parameters, the compressibility of the crystal lattice, the coefficients of thermal expansion, and the vibrational spectra were calculated in a series of classical molecular dynamics simulations using the ClayFF-MOH model and compared with a diverse set of available experimental data, including X-ray diffractometry, neutron scattering, IR and Raman spectroscopy. These new results demonstrated that ClayFF-MOH, as simple and approximate as it is, can be quite accurate in predicting many mineral properties at subduction zone conditions, which greatly expands the area of its applicability.

Keywords: brucite; atomistic simulations; high pressures; high temperatures; structural properties; elastic properties; bulk modulus; vibrational spectra

Citation: Tararushkin, E.V.; Pisarev, V.V.; Kalinichev, A.G. Equation of State, Compressibility, and Vibrational Properties of Brucite over Wide Pressure and Temperature Ranges: Atomistic Computer Simulations with the Modified ClayFF Classical Force Field. *Minerals* **2023**, *13*, 408. https://doi.org/10.3390/min13030408

Academic Editors: Lidong Dai, Haiying Hu and Jianjun Jiang

Received: 14 February 2023
Revised: 12 March 2023
Accepted: 13 March 2023
Published: 15 March 2023

1. Introduction

Brucite is a typical layered hydroxide mineral, $M(OH)_2$, where M is a divalent metal (M = Mg, Ca, Fe, Ni, etc.). Most such hydroxides relate to the layered brucite-type structure [1]. At ambient conditions, brucite crystal structure has a space group $P\bar{3}m1$ symmetry with layers consisting of edge-sharing $[MgO_6]$ octahedra. Hydrogen atoms attached to each oxygen form hydroxyl groups located above and below the octahedral layer and oriented along the *c*-axis. Despite the fact that the local environment of the OH groups appears to be quite unfavorable for hydrogen bonding between the layers, most authors argue that the nature of the interlayer interactions is due precisely to the hydrogen bonding [2].

The thermodynamic and elastic properties of brucite, $Mg(OH)_2$, especially at high pressures and temperatures, are of great importance in mineralogy and geophysics because brucite itself plays a major role in mineral equilibria controlling water balance in the subduction zones of the Earth's mantle, but also because the octahedral $Mg(OH)_6$ structural units and brucite-type layers constitute an important part of other minerals, such as layered dense magnesium hydrous silicates (DMHS) important in mantle mineralogy [3–6].

Therefore, the properties of brucite have been extensively studied experimentally by various techniques over wide ranges of temperature and pressure [2,7–9]. One of the most interesting crystallographic questions is the behavior of hydrogen (deuterium) atoms and the nature of hydrogen bonds under these conditions [8,10–13].

The structural and vibrational properties of brucite have been investigated in a number of experimental studies. In studying the structural parameters of brucite, depending on pressure and temperature, the positions of atoms and the distances between them are of interest [11,12,14–16] as well as the unit cell parameters (a, c, c/a, V) [7–9,11,12,14–19]. Such structural parameters of brucite were obtained using powder X-ray diffraction [7–9,15,16,18], single-crystal X-ray diffraction [17] and powder neutron diffraction [20,21]. The mechanical parameters of brucite are also of great interest, such as isothermal compressibility, bulk modulus, elastic constants, and thermal expansion coefficient [8,9,15–17,19–22]. Vibration properties of brucite under high pressure and temperature were also investigated in situ by Raman spectroscopy [2] and infrared spectroscopy [1,9,23–25]. The most attention has been paid to the study of OH librations, OH stretching, and the lattice vibrations at various values of pressure and temperature.

In addition to experimental studies, the structure, and vibrational properties of brucite were also evaluated by first-principles molecular dynamics simulations based on the density functional theory (DFT) and by classical molecular dynamic (MD) simulations [2,6,10,12,13,26–34]. Such simulations allow data to be obtained on the crystal lattice parameters, interatomic distances (O–H, O–Mg), elastic constants and bulk modulus, the coefficients of compressibility and thermal expansion of the brucite crystals, as well as their vibrational spectra. In addition, DFT modeling allows for the quantitative probing of the proton disorder [12,30] and dispersive interactions [13] in layered materials. For classical MD simulations of brucite, the ClayFF force field is most frequently used [26–28]. ClayFF is a nonbonded force field for classical atomistic simulations of a wide range of clay-related and cement-related materials [31–37]. In addition to modeling bulk mineral phases, ClayFF is also well suited for MD simulations of the interaction of water and aqueous solutions with mineral surfaces [26,28,31–37]. This force field was recently upgraded by adding several new terms that allow for a more accurate description of the M–O–H angular bending in hydroxide crystals [33,34,38].

In this study, we used this modified version of ClayFF to test its applicability in atomistic simulations of layered hydroxide minerals at high temperatures and high pressures by simulating the structural and vibrational properties of brucite. The atomistic model of brucite was isothermally compressed to 19.86 GPa at 300 K, to 15.15 GPa at 473 K, and to 14.56 GPa at 673 K. The simulation results were then carefully compared with all available experimental data and with the simulation results with the original version of ClayFF force field [31]. These new results clearly demonstrated that the ClayFF-MOH model can quite accurately predict many mineral properties under conditions well beyond its original range of parametrization [31]. This conclusion greatly expands the area of the applicability for this simple model, in particular for atomistic simulations of minerals and mineral-water interactions at subduction zone conditions.

2. Models and Methods

2.1. Structural Models

A brucite crystal model was constructed based on the structural data from powder neutron diffraction measurements at room temperature and pressure with the lattice parameters corresponding to a trigonal, $P\bar{3}m1$, unit cell ($a = b = 3.14979$ Å; $c = 4.7702$ Å; $\alpha = \beta = 90°$; $\gamma = 120°$) [21]. This unit cell was used as the initial building block for a simulation supercell containing 968 units ($11 \times 11 \times 8$ along the a, b, and c crystallographic directions, respectively). As a result, the simulation supercell contained 4840 atoms and had total dimensions of $34.65 \times 34.65 \times 38.16$ Å^3 (see Figure S1 in the Supplementary Materials).

Periodic boundary conditions were then applied in all three dimensions of the supercell to produce a continuous model of the crystal [39].

2.2. Force Field Parameters

All interatomic interactions were calculated using the ClayFF force field [1,34], which has already proven to be highly effective in modeling the structures of many oxides, hydroxides, clays, and other layered materials [33–37]. One of the objectives of the present work was to compare the results obtained by the original ClayFF version [1], further referred to as ClayFF-orig, and its recently developed modification, ClayFF-MOH, [31,33], which more accurately accounts for the bending of M–O–H angles of the hydroxide groups in the crystal by introducing an additional harmonic term to the force field expression:

$$E_{MOH} = k(\theta - \theta_0)^2 \tag{1}$$

where $\theta_{0,MgOH} = 110°$ and $k_{MgOH} = 6$ kcal·mol^{-1}·rad^{-2} were used for brucite [33]. For a more accurate modeling of the vibrational properties of brucite, the Morse potential was used to describe the hydroxide O–H stretching vibrations instead the original harmonic potential [28,40]:

$$E_{OH} = D_0\left[1 - e^{-\alpha(r - r_0)}\right]^2 \tag{2}$$

where $D_0 = 132.2491$ kcal·mol^{-1}, $\alpha = 2.1350$ Å^{-1}, $r_0 = 0.9572$ Å.

However, the primary objective of the present work was to test the temperature and pressure limits of the ClayFF applicability for the modeling of mineral properties by classical atomistic simulations at thermodynamic conditions beyond the range of original ClayFF parametrization [31].

2.3. Simulation Details

LAMMPS simulation package [41,42] was used to perform the MD simulations. As the first step, the potential energy of the brucite model supercell was minimized. Then an MD simulation run in *NPT*-ensemble was used to bring the supercell to equilibrium at the desired temperature and pressure. The Nosé–Hoover algorithm [39] was employed to control the temperature and pressure in the *NPT*-simulations. The *NPT*-simulation run lasted for 0.5 ns using the velocity Verlet's algorithm [39], which then followed by the *NVT*-simulation run for another 0.5 ns. The *NVT*-simulation was used to collect the equilibrium dynamic trajectories of all atoms for further statistical analyses. The coordinates and velocities of all atoms were recorded every 100 fs, which is enough to guarantee the statistical stability of data analysis. A timestep of 1.0 fs was always used for numerical integration of the equations of atomic motions.

2.4. Simulation Analysis

Structural and vibrational properties were evaluated in this study by time averaging over the entire simulated equilibrium MD trajectories. In particular, the lattice parameters of the bricute unit cell (a, c, c/a, V) and volume compression of brucite, isothermal bulk moduli, power spectra of vibrational density of states were calculated, as well as maps of atomic density and angular distributions for hydroxyls in the octahedral sheets of brucite.

The unit cell parameters of brucite were directly calculated as averages from the equilibrium *NPT*-simulations. The isothermal volume compression of the unit cell, V/V_0, was also calculated for the corresponding pressures from the equilibrium *NPT*-simulations.

The isothermal bulk modulus and its pressure derivative were calculated by least squares fitting the third-order Birch–Murnaghan equation of state in the Eulerian finite strain parameter [43,44] to the simulated pressure–volume data:

$$P = \frac{3}{2}K_{T0}\left[\left(\frac{V_0}{V}\right)^{\frac{7}{3}} - \left(\frac{V_0}{V}\right)^{\frac{5}{3}}\right]\cdot\left\{1 + \frac{3}{4}(K'_{T0} - 4)\cdot\left[\left(\frac{V_0}{V}\right)^{\frac{2}{3}} - 1\right]\right\} \tag{3}$$

where K_{T0} and K'_{T0} are the isothermal bulk elastic modulus and its pressure derivative, respectively.

The volumetric thermal expansion coefficient of brucite was obtained by fitting the *P*-*V*-*T* simulation data via constraints on the thermoelastic properties at high-pressure [15,45]. The third-order Birch–Murnaghan equation of state was used to calculate the thermal expansion coefficient through the fitting to a liner relationship [15]:

$$\alpha(T) = \alpha_0 + \alpha_1 T \quad (4)$$

The amplitudes of atomic vibrations in the brucite crystal were visualized by calculating the time-averaged contour maps of atomic densities for a slice of the crystal starting from the positions of Mg atoms to the positions of H atoms of the corresponding hydroxyl groups. The thickness of this slice was ~3.0 Å. The atomic density of an atom α, $\rho_\alpha(x,y)$, was evaluated on the square grid and the time-averaged number of atoms $\langle N_\alpha(\Delta x \Delta y)\rangle$ in each grid cell normalized by the surface area of the cell:

$$\rho_\alpha(x,y) = \frac{\langle N_\alpha(\Delta x \Delta y)\rangle}{\Delta x \Delta y} \quad (5)$$

For geometrical analyses of the hydroxyl (O–H bond) angular distributions the angles θ and φ were calculated for both versions of the ClayFF force field, as defined in Figure 1.

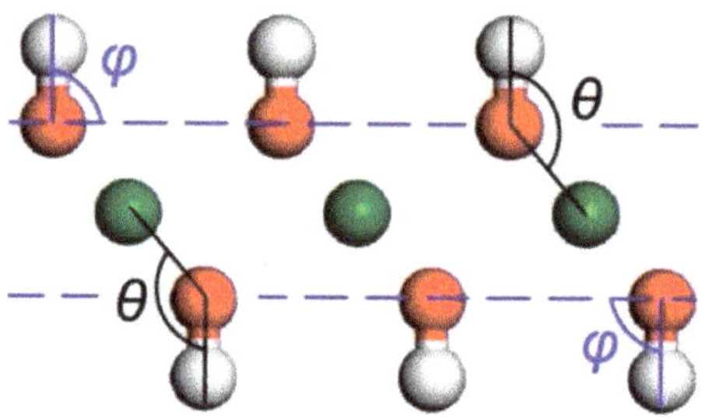

Figure 1. Angle of slope, φ, of the structural hydroxyls to the (001) plane of the brucite crystal and the Mg–O–H angle, θ. Green atoms—Mg, red atoms—O, white atoms—H.

The power spectra (PS), providing the quantitative information about the frequencies of atomic vibrations in the system, were obtained as Fourier transformations of the velocity autocorrelation function (VACF) of the corresponding atoms (e.g., [27,37,46,47]):

$$PS(\omega) = \sum_{t=1}^{n_{\text{VACF}}} \text{VACF}(t) \cdot \cos\left(\frac{\pi}{n_{\text{VACF}}} \cdot t \cdot \omega\right) \quad (6)$$

where n_{VACF} is the number of trajectory 'windows' used in a particular VACF calculation. The normalized VACFs were calculated from the MD-simulated equilibrium *NVT* trajectories of the atoms in the modeled system [47]:

$$\text{VACF}(t) = \frac{\sum_{f=1}^{n-n_{\text{VACF}}} \sum_{sel.atoms} \vec{v}_f \cdot \vec{v}_{f+t}}{\sum_{f=1}^{n-n_{\text{VACF}}} \sum_{sel.atoms} \vec{v}_f \cdot \vec{v}_f} \quad (7)$$

where n is the total number of the trajectory 'windows'; v_f and v_{f+t} are velocities at times f and $f + t$, respectively. They were smoothed by applying a filtering function [47]:

$$\text{VACF}(t) = \text{VACF}(t) \cdot exp\left(-\frac{t}{\tau}\right) \quad (8)$$

The VACF and PS were calculated for all atoms of the crystal and separately for each atom type (magnesium, oxygen, and hydrogen). We also analyzed the *xy* and *zz* components of the VACF tensor to obtain PS in the *xy* direction (parallel to the plane of brucite layers) and *zz* direction (perpendicular to the layering).

3. Results and Discussion

3.1. Compressibility and Lattice Parameters

First, we checked how the original and modified versions of the ClayFF force field reproduce the elastic properties of brucite. The compressibility of brucite was studied at temperatures 300, 473, and 673 K and compared with the experimental data [8,9]. For the ClayFF-orig model, the calculations were carried out only up to $T = 473$ K, because at $T = 673$ K, the brucite supercell became already unstable. For the ClayFF-MOH model, the supercell remains stable up to $T = 673$ K, but even with that modification the supercell loses stability at $T = 873$ K.

The compressibility was calculated at pressures up to 20 GPa at $T = 300$ K and up to 15 GPa at 473 K and 673 K. The experimental data and the calculated values of compressibility at $T = 300$ K shown in Figure 2. In this case, good agreement with experimental data is observed in the entire pressure range. A slightly larger deviation is observed for the ClayFF-MOH force field for higher pressures of ~20 GPa. The compressibility of brucite at temperatures of 473 K and 673 K is shown in Figure 2. At 473 K, a good agreement with the experimental data is found again, and the ClayFF-MOH force field results in slightly larger deviations from the experiment. However, only the ClayFF-MOH results are available for $T = 673$ K, as the brucite supercell decomposes with the ClayFF-orig simulations at that temperature even at normal pressure.

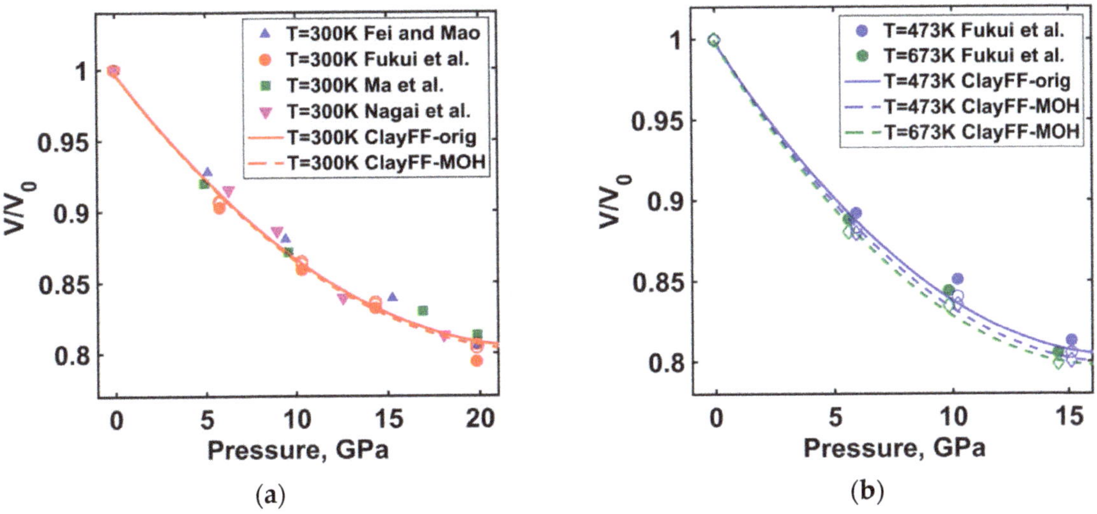

Figure 2. Compressibility of brucite: (**a**) $T = 300$ K; (**b**) $T = 473$ K and $T = 673$ K. Open circles and rhombs belong to calculated values. Experimental data are from Fei and Mao [7], Fukui et al. [8], Ma et al. [9], and Nagai et al. [16].

A comparison of the crystallographic parameters of brucite at high temperatures and high pressures with experimental data [8] show quite a good agreement (see Figure S2 in the Supplementary Materials). The lattice parameters of the unit cell for ambient conditions calculated in this work virtually coincide with the lattice parameters reported by Pouvreau et al. [33] for the ClayFF-orig version, while for the ClayFF-MOH version of the force field our simulations results ($a = 3.29$ Å, $c = 4.77$ Å, $V = 44.12$ Å^3) are slightly higher.

The maximum deviation of the unit cell volume from experimental values is 9% at the highest pressures for both versions of the ClayFF force field at temperatures 300 and 473 K. At T = 673 K, the maximum deviation of the unit cell volume for the ClayFF-MOH model is 11%.

The agreement of the parameter a of the unit cell with experiment [8] is fairly good, with the discrepancy from the experimental value no more than 5%. There is also a good agreement in the uniaxial compression curve along the a axis, including the high-temperature calculations.

For ambient conditions, the deviations from the experimental data [8] for the c axis are −1% and −2% for the ClayFF-orig and ClayFF-MOH versions, respectively. However, both versions of the force field do not reproduce the c axis temperature dependence. At high pressures, the discrepancy reaches up to 5%. This leads to an incorrect trend of the c/a ratio dependence on pressure (Figure S2 in the Supplementary Materials).

3.2. Bulk Modulus

The values of bulk modulus K_{T0} and its pressure derivative K'_{T0} computed from Equation (3) are shown in Figure 3 in comparison with the experimental data of Fukui et al. [8]. Other experimental data at room temperature, e.g., [15,17,18,21] and the data of DFT(GGA) calculations, e.g., [10,12], are not shown in Figure 3, as they are very close to the data of Fukui et al. [8]. The calculated bulk elastic modulus at temperatures of 300 K and 473 K for both versions of the ClayFF force field shows a good agreement with the experimental data [8], and the maximum discrepancy is only ~3% for the ClayFF-MOH model. For a higher temperature of 673 K, the deviation of the bulk modulus from the experiment for the ClayFF-MOH force field is ~11%, which is also not bad.

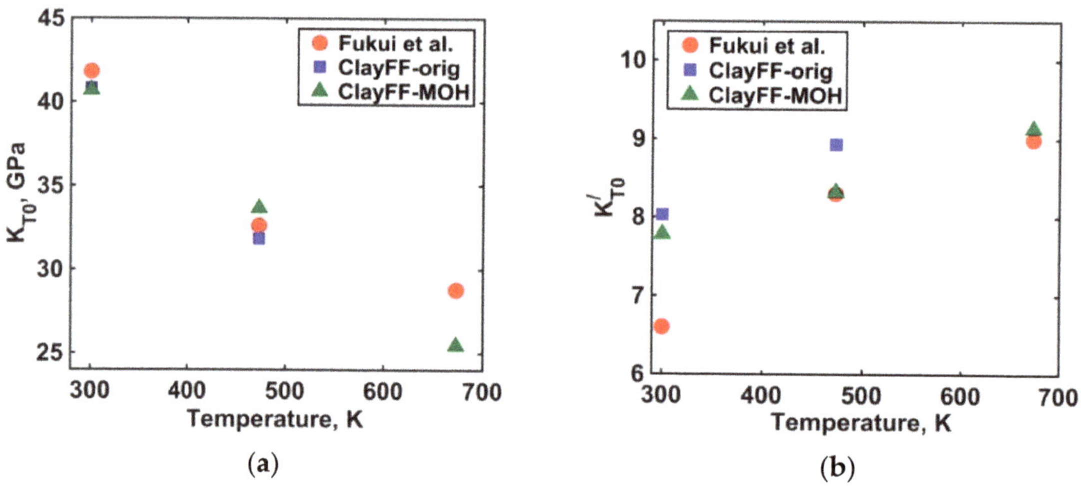

Figure 3. (**a**) Bulk elastic modulus of brucite; and (**b**) pressure derivative of the bulk modulus. Experimental data of Fukui et al. [8] are shown for comparison.

The pressure derivatives of the bulk modulus at T = 300 K gave the largest deviation from the experimental data, with the maximum difference of ~21% for the ClayFF-orig model. In fact, the calculated pressure derivatives at room temperature closely agree with the pressure derivatives measured by Catty et al. [21], the largest difference being ~6% for the ClayFF-orig model. With increasing temperature, the first derivative gives a better agreement with experiment [8], and for a T = 673 K the difference is ~2% for the ClayFF-MOH model.

3.3. Thermal Expansion Coefficient

The values of the two coefficients of thermal expansion (α_0 and α_1 in Equation (4)) were only calculated from the MD results obtained for the ClayFF-MOH model due to the instability of the brucite supercell at T = 673 K for the ClayFF-orig model. The values of the coefficients for the ClayFF-MOH model are $\alpha_0 = 9.04 \times 10^{-5}$ K^{-1} and $\alpha_1 = 13.00 \times 10^{-8}$ K^{-2}. The value of α_0 is close to the experimentally determined value of α_0 = 7.3(3) $\times 10^{-5}$ K^{-1} [15], while the value of α_1 differs more from the experiment (α_1 = 3.6(13) $\times 10^{-8}$ K^{-2} [15]).

3.4. Structural Parameters

Nearest-neighbor Mg–O distances were calculated for both ClayFF versions (Figure S3a). At ambient conditions, both versions of the force field give slightly larger Mg–O distances (2.2–2.3 Å) compared to the experimental [21] and DFT [28] results (2.12–2.13 Å). With increasing pressure, at T = 300 K, the Mg–O distance reduces, as well as its mean-squared deviation. Overall, the agreement in the Mg–O distances between the ClayFF models and experimental data [16,21] even improves at higher pressures. The same trends of decreasing Mg–O distance and its fluctuations with increasing pressure are maintained at higher temperatures (473 and 673 K). The distances obtained by both ClayFF models virtually coincide, the difference is only in a larger fluctuation for the ClayFF-orig version, which indicates a more stable brucite structure when the M–O–H angles in the ClayFF-MOH model are more accurately constrained.

The distributions and averages of the hydroxyl O–H distances in brucite for both versions of the ClayFF force field were also calculated. At ambient pressure, the O–H distance is close to the O–H bond length corrected for thermal motion [21] and the bond lengths calculated by DFT [10]. The average O–H distance slightly increases with increasing pressure, which does not agree with experimental data [21] but agrees well with the DFT results at high pressures [10]. The tendency to increase the O–H distance with increasing pressure persists at elevated temperatures. Similar to the Mg–O distance, increasing temperature leads to a slight increase of the average O–H bond length and its dispersion. The lack of good agreement with these experimental data at elevated pressures can be explained by the behavior of the Morse potential used here for modeling the O–H bond of hydroxyls, which has a fairly strong repulsive branch of the potential when hydrogen atom approaches the oxygen. On the other hand, an increase in O–D distance with increasing pressures is indeed experimentally observed for deuterated brucite [20]. So, a better judgement about the quality of the description of this parameter with the ClayFF model could probably be made when more experimental data are available. All the comparisons are shown in Figure S3b in the Supplementary Materials.

The angular distribution of the hydroxyl orientations relative to the brucite (001) plane of layering at T = 300 K and high pressures for both versions of the ClayFF model are shown in Figure 4. These distributions are qualitatively different between the two ClayFF versions. At ambient pressure, the direction normal to the brucite layers is the most probable hydroxyl orientation. However, the distribution is significantly broader for the ClayFF-orig model where hydroxyl orientations are not constrained and have the ability to lean toward the brucite layer. With increasing pressure, the orientation normal to the layering remains the most probable for the ClayFF-MOH model, but the earlier ClayFF-orig model demonstrates the appearance of another probability maximum. For example, the most probable orientation is ~63° at P = 10.35 GPa. Small deviations of hydroxyl orientations from the normal to the brucite surfaces at ambient pressure is in good agreement with the angles calculated from experimental data [48]. According to experimental data [21], angle φ (see Figure 1) deviates from the normal by only 6.8° at P = 10.9 GPa, compared to the same angle under ambient conditions, which, in turn, indicates that hydroxyls remain nearly normal to the layering even at very high pressures, so that the angular distributions for the ClayFF-orig model exhibit unphysical behavior under these conditions. At higher temperatures, in all cases, the overall shape of the hydroxyl orientational distribution remains the same (Figure 4b), but a broadening is observed, consistent with higher thermal energies of angular vibrations at higher temperatures.

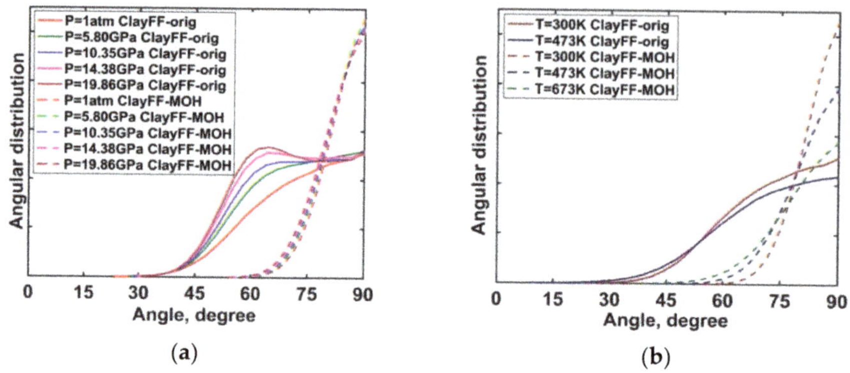

Figure 4. (**a**) Angular distributions of hydroxyl orientations at T = 300 K and high pressures; and (**b**) angular distributions of hydroxyl orientations at T = 473 K and T = 673 K at P = 1 bar.

The nature of the second maximum at ~60° for the ClayFF-orig model can be understood by considering the distribution of atomic densities projected onto the brucite *ab* crystallographic plane. Atomic density maps for brucite layer at T = 473 K and P = 15.15 GPa for both ClayFF versions are shown in Figure 5.

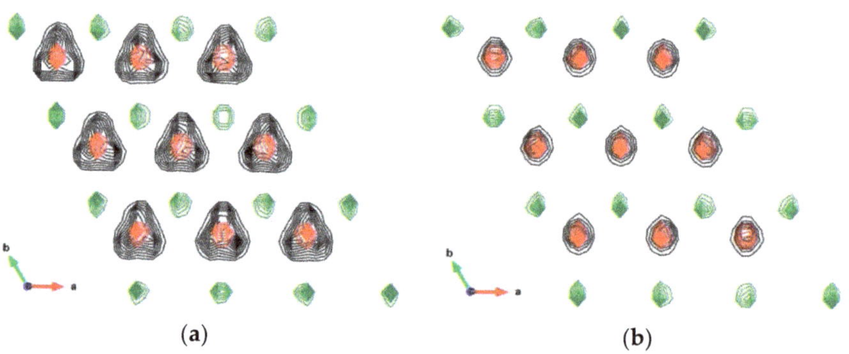

Figure 5. Contour maps of atomic densities at T = 473 K and P = 15.15 GPa for 3 × 3 unit cell. (**a**) ClayFF-orig; and (**b**) ClayFF-MOH. Green contours—Mg, red contours—O, black contours—H.

The ClayFF-orig model shows three predominant orientations of the O–H groups forming an equilateral triangle of probability distributions. These orientations are not accidental, they emerge because the H atoms of the unconstrained O–H groups can easily form hydrogen bonds with one of the three O atoms of the opposite brucite layer. For the ClayFF-MOH model, the atomic density maps look quite different, the O–H groups are now constrained in their orientations and their H atoms are located more or less above the respective O atoms. In this case, hydrogen bonds with the opposite layer are also formed, but instead of one strong bond towards one of the opposing O atoms, three weaker bonds with three nearest O atoms exist simultaneously. In other words, due to the energy penalty from the new M–O–H angle bending terms of the force field, this interaction is not chaotic and H atom stays near the center of a triangle formed by the closest opposite-layer O atoms.

3.5. Power Spectra

The spectra of vibrational density of states, or so-called power spectra (Equations (6)–(8)) were calculated for both ClayFF versions and shown in Figure 6. Experimental studies [1,2,9,14,23–25] and atomistic simulations [6,10,13,27,28] indicate that the most intense vibrational bands of brucite under ambient conditions are in the frequency ranges of 270–900 cm^{-1} and 3550–3890 cm^{-1}. The lower frequency bands correspond to the O–H librational modes (E_g and E_u) and to the lattice vibrations (A_{1g}, A_{2u}, E_g and E_u). The higher frequency bands correspond to the O–H stretching modes (A_{1g} and A_{2u}) [1,2,9,10,13,14,25,27].

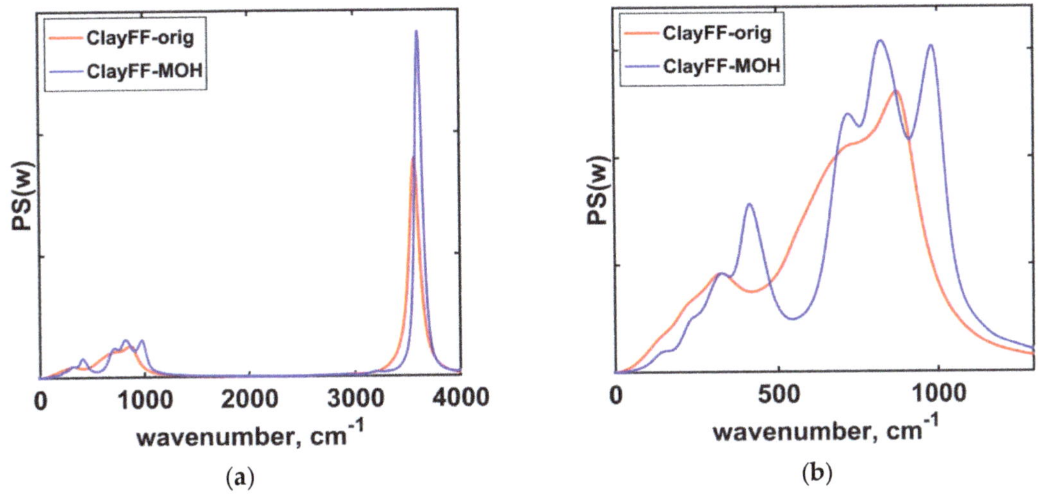

Figure 6. Power spectra of brucite at T = 298 K and P = 1 bar: (**a**) full spectrum; and (**b**) lower frequency modes.

The calculated power spectra of brucite under ambient conditions are in good agreement with the previous results [28] for the ClayFF force field with the constrained Mg–O–H angles. Zeitler et al. [28] have already conducted a detailed analysis of the peak positions in the low-frequency range of the spectrum. In particular, they have shown that taking into account the Mg–O–H angle allows for a slightly better agreement between the calculated classical MD and DFT results for the O–H librational modes, as it was also later confirmed by Pouvreau et al. [33]. Our decomposition of the total vibrational spectra of brucite into the contributions due to O and H atomic motions in the xy plane also shows the difference in the O–H librational modes between the two ClayFF versions (Figure S4 in the Supplementary Materials).

Our calculations show the following features of the spectral peaks in the lower frequency range at high pressures and temperatures:

- most peaks exhibit a red shift, and their intensity decreases with increasing temperature (Figure S5 in the Supplementary Materials);
- the peaks exhibit blue shift, and their intensity decreases with increasing pressure (Figure S6 in the Supplementary Materials).

Such tendencies at high pressures agree with the experimental data [2]. Also, at high pressures there is a merging and splitting of some peaks (Figure S6 in the Supplementary Materials).

The higher frequency range of the spectra describes the O–H bond-stretching modes, which correspond to the modes A_{1g} (Raman) and A_{2u} (infrared) in experiments. For these modes, the peaks are located at 3573 cm^{-1} and 3612 cm^{-1} for the ClayFF-orig and ClayFF-MOH models, respectively (Figure 6). As the frequency is mainly determined by the O–H bond parameters, the peak positions are in good agreement with experimental studies and

DFT calculations [1,2,6,9,10,13–15,24,27] for both ClayFF versions. However, the ClayFF versions differ in the width and intensity of the peaks: the O–H stretching peak is narrower and more intense for the ClayFF-MOH, compared to ClayFF-orig model.

At ambient pressure, the O–H stretching peak does not change its position with temperature in the ClayFF-MOH model. The increasing temperature broadens the peak and reduces its intensity due to the thermal motion (Figure S5 in the Supplementary Materials). However, the peaks shift with the increasing pressure indicating a change in the effective bond stiffness due to the interaction with the neighboring atoms. The computed O–H stretching peak positions for both versions of the ClayFF force field are shown in Figure 7 together with Raman experimental data and DFT results. One can observe that the change of the peak position is smaller for ClayFF-MOH than for ClayFF-orig. Such a difference is likely caused by the difference in the orientational dynamics of the O–H bonds between the two force field versions: as discussed in Section 3.4, the O–H bonds in ClayFF-orig are inclined towards a specific oxygen atom, while in ClayFF-MOH they stay more or less normal to the surface. That difference in the local atomic environments produces the difference in the stiffness of the O–H groups, in addition to the bond potential. Overall, ClayFF-MOH shows a better agreement with the experimental data [2] and DFT calculations [10], especially at the highest pressures (~15 GPa and ~20 GPa), showing that the constrained O–H orientation leads to a better description of the vibrational dynamics. The above-described behavior of the peaks at high pressures persists at elevated temperatures as well (Figure S7 in the Supplementary Materials).

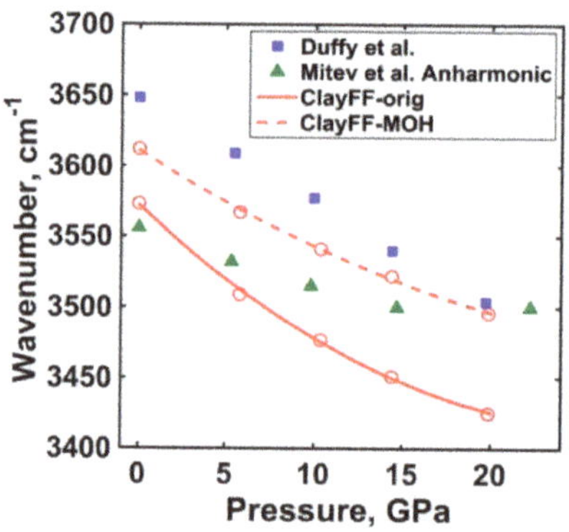

Figure 7. Calculated and experimental O–H stretching peak positions at high pressures and room temperature. Experimental data [2] and DFT results [10] are shown for comparison.

4. Conclusions

The properties of brucite under ambient conditions and up to high temperature and high-pressure conditions were investigated by classical MD simulations using the original ClayFF force field [31] and its more recent modification ClayFF-MOH [34], where the bending motions of the Mg–O–H angles are more accurately constrained by additional terms based on the DFT calculations [33]. For both models, the crystallographic, elastic, and vibrational properties acceptably agreed with available experimental data and the results of more rigorous DFT calculations. However, the introduction of the Mg–O–H angle bending term made it possible to simulate brucite at higher temperatures and pressures (T = 673 K and P = 15 GPa) than with the original ClayFF parameterization. The isothermal compressibility of brucite was also reproduced well by both versions of the force field, including at high temperatures. The bulk modulus was also well reproduced, while at higher temperatures the agreement with the experimental data worsened. The crystal lattice unit cell parameters (a, c, V and c/a) were reproduced with acceptable accuracy. In most cases (a, c and V), the trends of the lattice parameters' dependence on pressure and temperature for the ClayFF model agreed well with the experimental data and DFT results.

The application of the new more accurate ClayFF-MOH force field results in the formation of a more realistic hydrogen bonding structure between brucite layers, which leads to a more stable brucite structure at higher temperatures and pressures, and a more realistic description of its vibrational spectrum.

It is important to emphasize that both versions of the ClayFF model were intentionally used here well beyond the range of temperatures and pressures for which this model was initially parameterized [31,33,34]. Therefore, the results of this work are quite encouraging in view of the applicability of these simple models to the classical atomistic simulations of more complex hydrous minerals under the subduction zone conditions (e.g., [49,50]). This will be the focus of our further investigation.

Supplementary Materials: The following supporting information can be downloaded at: https://www.mdpi.com/article/10.3390/min13030408/s1. Figure S1: Two projections of the brucite simulation supercell; Figure S2: (a) Volume of the unit cell for brucite as function of pressure; (b) pressure dependence of the parameter *a* of brucite; (c) pressure dependence of the parameter *c* for brucite; (d) ratio of unit cell parameters *c*/*a* for brucite as a function of pressure; Figure S3: (a) Pressure dependence of the simulated Mg–O distances in $[MgO_6]$ octahedra; (b) pressure dependence of the simulated O–H distances in structural hydroxyls; Figure S4: Power spectra for H_{xy} (a) and O_{xy} (b) for both versions of the ClayFF force field; Figure S5: Power spectra of brucite at $P = 1$ bar and high temperatures for the ClayFF-MOH model; Figure S6: Power spectra of brucite at $T = 300$ K and high pressures for the ClayFF-MOH model; Figure S7: Power spectra of brucite at $T = 673$ K and high pressures for the ClayFF-MOH model.

Author Contributions: Conceptualization, A.G.K.; methodology, A.G.K. and V.V.P.; investigation, E.V.T.; data analysis E.V.T., V.V.P. and A.G.K.; writing—original draft preparation, E.V.T.; writing—review and editing, E.V.T., V.V.P. and A.G.K.; visualization, E.V.T.; supervision, V.V.P. and A.G.K.; project administration, V.V.P. and A.G.K.; funding acquisition, E.V.T., V.V.P. and A.G.K. All authors have read and agreed to the published version of the manuscript.

Funding: This study was sponsored by Agence Nationale pour la Gestion des Déchets Radioactifs and National Research University Higher School of Economics.

Data Availability Statement: All data are available from the authors upon request.

Acknowledgments: This research was funded by the HSE University Basic Research Program and was supported by computational resources of HPC facilities at HSE University [51]. A.G.K. also acknowledges the financial support of the industrial chair "Storage and Disposal of Radioactive Waste" at the IMT Atlantique, funded by ANDRA, Orano, and EDF. The APC was funded by publication vouchers from previous MDPI reviewing activity.

Conflicts of Interest: The authors declare no conflict of interest.

References

1. Lutz, H.D.; Moller, H.; Schmidt, M. Lattice vibration spectra. Part LXXXII. Brucite-type hydroxides $M(OH)_2$ (M = Ca, Mn, Co, Fe, Cd)—IR and Raman spectra, neutron diffraction of $Fe(OH)_2$. *J. Mol. Struct.* **1994**, *328*, 121–132. [CrossRef]
2. Duffy, T.S.; Meade, C.; Fei, Y.; Mao, H.-K.; Hemley, R.J. High-pressure phase transition in brucite, $Mg(OH)_2$. *Am. Mineral.* **1995**, *80*, 222–230. [CrossRef]
3. Thompson, A.B. Water in the Earth's upper mantle. *Nature* **1992**, *358*, 295–302. [CrossRef]
4. Kawamoto, T. Hydrous phases and water transport in the subducting slab. *Rev. Mineral. Geochem.* **2006**, *62*, 273–289. [CrossRef]
5. Ivanov, A.V.; Litasov, K.D. The deep water cycle and flood basalt volcanism. *Int. Geol. Rev.* **2013**, *56*, 1–14. [CrossRef]
6. Hermann, A.; Mookherjee, M. High-pressure phase of brucite stable at Earth's mantle transition zone and lower mantle conditions. *Proc. Nat. Acad. Sci. USA* **2016**, *113*, 13971–13976. [CrossRef]
7. Fei, Y.; Mao, H.-K. Static compression of $Mg(OH)_2$ to 78 GPa at high temperature and constraints on the equation of state of fluid H_2O. *J. Geophys. Res. Solid Earth* **1993**, *98*, 11875–11884. [CrossRef]
8. Fukui, H.; Ohtaka, O.; Suzuki, T.; Funakoshi, K. Thermal expansion of $Mg(OH)_2$ brucite under high pressure and pressure dependence of entropy. *Phys. Chem. Miner.* **2003**, *30*, 511–516. [CrossRef]
9. Ma, M.; Liu, W.; Chen, Z.; Liu, Z.; Li, B. Compression and structure of brucite to 31 GPa from synchrotron X-ray diffraction and infrared spectroscopy studies. *Am. Mineral.* **2013**, *98*, 33–40. [CrossRef]
10. Mitev, P.D.; Gajewski, G.; Hermansson, K. Anharmonic OH vibrations in brucite: Small pressure-induced redshift in the range 0–22 GPa. *Am. Mineral.* **2009**, *94*, 1687–1697. [CrossRef]

11. Xu, H.; Zhao, Y.; Hickmott, D.D.; Lane, N.J.; Vogel, S.C.; Zhang, J.; Daemen, L.L. High-temperature neutron diffraction study of deuterated brucite. *Phys. Chem. Miner.* **2013**, *40*, 799–810. [CrossRef]
12. Mookherjee, M.; Stixrude, L. High-pressure proton disorder in brucite. *Am. Mineral.* **2006**, *91*, 127–134. [CrossRef]
13. Ugliengo, P.; Zicovich-Wilson, C.M.; Tosoni, S.; Civalleri, B. Role of dispersive interactions in layered materials: A periodic B3LYP and B3LYP-D* study of $Mg(OH)_2$, $Ca(OH)_2$ and kaolinite. *J. Mater. Chem.* **2009**, *19*, 2564–2572. [CrossRef]
14. Chakoumakos, B.C.; Loong, C.K.; Schultz, A.J. Low-temperature structure and dynamics of brucite. *J. Phys. Chem. B* **1997**, *101*, 9458–9462. [CrossRef]
15. Xia, X.; Weidner, D.J.; Zhao, H. Equation of state of brucite; single-crystal Brillouin spectroscopy study and polycrystalline pressure-volume-temperature measurement. *Am. Mineral.* **1998**, *83*, 68–74. [CrossRef]
16. Nagai, T.; Ito, T.; Hattori, T.; Yamanaka, T. Compression mechanism and amorphization of portlandite, $Ca(OH)_2$: Structural refinement under pressure. *Phys. Chem. Miner.* **2000**, *27*, 462–466. [CrossRef]
17. Duffy, T.S.; Shu, J.; Mao, H.-K.; Hemley, R.J. Single-crystal X-ray diffraction of brucite to 14 GPa. *Phys. Chem. Miner.* **1995**, *22*, 277–281. [CrossRef]
18. Horita, J.; dos Santos, A.M.; Tulk, C.A.; Chakoumakos, B.C.; Polyakov, V.B. High-pressure neutron diffraction study on H-D isotope effects in brucite. *Phys. Chem. Miner.* **2010**, *37*, 741–749. [CrossRef]
19. Redfern, S.A.T.; Wood, B.J. Thermal expansion of brucite, $Mg(OH)_2$. *Am. Mineral.* **1992**, *77*, 1129–1132.
20. Parise, J.B.; Leinenweber, K.; Weidner, D.J.; Tan, K.; Von Dreele, R.B. Pressure-induced H bonding: Neutron diffraction study of brucite, $Mg(OD)_2$, to 9.3 GPa. *Am. Mineral.* **1994**, *79*, 193–196.
21. Catti, M.; Ferraris, G.; Hull, S.; Pavese, A. Static compression and H disorder in brucite, $Mg(OH)_2$, to 11 GPa: A powder neutron diffraction study. *Phys. Chem. Miner.* **1995**, *22*, 200–206. [CrossRef]
22. Jiang, F.; Speziale, S.; Duffy, T.S. Single-crystal elasticity of brucite, $Mg(OH)_2$, to 15 GPa by Brillouin scattering. *Am. Mineral.* **2006**, *91*, 1893–1900. [CrossRef]
23. Kruger, M.B.; Williams, Q.; Jeanloz, R. Vibrational spectra of $Mg(OH)_2$ and $Ca(OH)_2$ under pressure. *J. Chem. Phys.* **1989**, *91*, 5910–5915. [CrossRef]
24. de Oliveira, E.F.; Hase, Y. Infrared study and isotopic effect of magnesium hydroxide. *Vib. Spectrosc.* **2001**, *25*, 53–56. [CrossRef]
25. Zhu, X.; Guo, X.; Smyth, J.R.; Ye, Y.; Wang, X.; Liu, D. High-temperature vibrational spectra between $Mg(OH)_2$ and $Mg(OD)_2$: Anharmonic contribution to thermodynamics and D/H fractionation for brucite. *J. Geophys. Res. Solid Earth* **2019**, *124*, 8267–8280. [CrossRef]
26. Wang, J.W.; Kalinichev, A.G.; Kirkpatrick, R.J. Molecular modeling of water structure in nano-pores between brucite (001) surfaces. *Geochim. Cosmochim. Acta* **2004**, *68*, 3351–3365. [CrossRef]
27. Braterman, P.S.; Cygan, R.T. Vibrational spectroscopy of brucite: A molecular simulation investigation. *Am. Mineral.* **2006**, *91*, 1188–1196. [CrossRef]
28. Zeitler, T.R.; Greathouse, J.A.; Gale, J.D.; Cygan, R.T. Vibrational analysis of brucite surfaces and the development of an improved force field for molecular simulation of interfaces. *J. Phys. Chem. C* **2014**, *118*, 7946–7953. [CrossRef]
29. Ulian, G.; Valdre, G. Equation of state and second-order elastic constants of portlandite $Ca(OH)_2$ and brucite $Mg(OH)_2$. *Phys. Chem. Miner.* **2018**, *46*, 101–117. [CrossRef]
30. Schaack, S.; Depondt, P.; Huppert, S.; Finocchi, F. Quantum driven proton diffusion in brucite-like minerals under high pressure. *Sci. Rep.* **2020**, *10*, 8123. [CrossRef]
31. Cygan, R.T.; Liang, J.-J.; Kalinichev, A.G. Molecular models of hydroxide, oxyhydroxide, and clay phases and the development of a general force field. *J. Phys. Chem. B* **2004**, *108*, 1255–1266. [CrossRef]
32. Cygan, R.T.; Greathouse, J.A.; Heinz, H.; Kalinichev, A.G. Molecular models and simulations of layered materials. *J. Mater. Chem.* **2009**, *19*, 2470–2481. [CrossRef]
33. Pouvreau, M.; Greathouse, J.A.; Cygan, R.T.; Kalinichev, A.G. Structure of hydrated gibbsite and brucite edge surfaces: DFT results and further development of the ClayFF classical force field with metal-O-H angle bending terms. *J. Phys. Chem. C* **2017**, *121*, 14757–14771. [CrossRef]
34. Cygan, R.T.; Greathouse, J.A.; Kalinichev, A.G. Advances in *clayff* molecular simulation of layered and nanoporous materials and their aqueous interfaces. *J. Phys. Chem. C* **2021**, *125*, 17573–17589. [CrossRef]
35. Kalinichev, A.G.; Wang, J.W.; Kirkpatrick, R.J. Molecular dynamics modeling of the structure, dynamics and energetics of mineral-water interfaces: Application to cement materials. *Cem. Concr. Res.* **2007**, *37*, 337–347. [CrossRef]
36. Mishra, R.K.; Mohamed, A.K.; Geissbühler, D.; Manzano, H.; Jamil, T.; Shahsavari, R.; Kalinichev, A.G.; Galmarini, S.; Tao, L.; Heinz, H.; et al. *cemff*: A force field database for cementitious materials including validations, applications and opportunities. *Cem. Concr. Res.* **2017**, *102*, 68–89. [CrossRef]
37. Kalinichev, A.G. Atomistic modeling of clays and related nanoporous materials with ClayFF force field. In *Computational Modeling in Clay Mineralogy*; Sainz-Díaz, C.I., Ed.; Association Internationale pour l'Etude des Argiles (AIPEA): Bari, Italy, 2021; Volume 3, pp. 17–52.
38. Pouvreau, M.; Greathouse, J.A.; Cygan, R.T.; Kalinichev, A.G. Structure of hydrated kaolinite edge surfaces: DFT results and further development of the ClayFF classical force field with Metal-O-H angle bending terms. *J. Phys. Chem. C* **2019**, *123*, 11628–11638. [CrossRef]
39. Allen, M.P.; Tildesley, D.J. *Computer Simulation of Liquids*, 2nd ed.; Oxford University Press: New York, NY, USA, 2017. [CrossRef]

40. Greathouse, J.A.; Durkin, J.S.; Larentzos, J.P.; Cygan, R.T. Implementation of a Morse potential to model hydroxyl behaviour in phyllosilicates. *J. Chem. Phys.* **2009**, *130*, 134713. [CrossRef]
41. Plimpton, S. Fast parallel algorithms for short-range molecular dynamics. *J. Comp. Phys.* **1995**, *117*, 1–19. [CrossRef]
42. Thompson, A.P.; Aktulga, H.M.; Berger, R.; Bolintineanu, D.S.; Brown, W.M.; Crozier, P.S.; in't Veld, P.J.; Kohlmeyer, A.; Moore, S.G.; Nguyen, T.D.; et al. LAMMPS–a flexible simulation tool for particle-based materials modeling at the atomic, meso, and continuum scales. *Comp. Phys. Comm.* **2022**, *271*, 108171. [CrossRef]
43. Birch, F. Finite strain isotherm and velocities for single-crystal and polycrystalline NaCl at high pressures and 300°K. *J. Geophys. Res. Solid Earth* **1978**, *83*, 1257–1268. [CrossRef]
44. Katsura, T.; Tange, Y. A simple derivation of the Birch-Murnaghan equations of state (EOSs) and comparison with EOSs derived from other definitions of finite strain. *Minerals* **2019**, *9*, 745. [CrossRef]
45. Jackson, I.; Rigden, S.M. Analysis of P-V-T data: Constraints on the thermoelastic properties of high-pressure minerals. *Phys. Earth Planet. Inter.* **1996**, *96*, 85–112. [CrossRef]
46. Kirkpatrick, R.J.; Kalinichev, A.G.; Wang, J.; Hou, X.; Amonette, J.E. Molecular modeling of the vibrational spectra of interlayer and surface species of layered double hydroxides. *CMS Workshop Lect.* **2005**, *13*, 239–285. [CrossRef]
47. Szczerba, M.; Kuligiewicz, A.; Derkowski, A.; Gionis, V.; Chryssikos, G.D.; Kalinichev, A.G. Structure and dynamics of water-smectite interfaces: Hydrogen bonding and the origin of the sharp O-DW/O-HW infrared band from molecular simulations. *Clays Clay Min.* **2016**, *64*, 452–471. [CrossRef]
48. Desgranges, L.; Calvarin, G.; Chevrier, G. Interlayer interactions in $M(OH)_2$: A neutron diffraction study of $Mg(OH)_2$. *Acta Crystallogr. Sect. B* **1996**, *52*, 82–86. [CrossRef]
49. Wang, J.; Kalinichev, A.G.; Kirkpatrick, R.J. Molecular modeling of the 10-Å phase at subduction zone conditions. *Earth Planet. Sci. Lett.* **2004**, *222*, 517–527. [CrossRef]
50. Wang, J.; Kalinichev, A.G.; Kirkpatrick, R.J. Structure and decompression melting of a novel, high-pressure nanoconfined 2-D ice. *J. Phys. Chem. B* **2005**, *109*, 14308–14313. [CrossRef]
51. Kostenetskiy, P.S.; Chulkevich, R.A.; Kozyrev, V.I. HPC Resources of the Higher School of Economics. *J. Phys. Conf. Ser.* **2021**, *1740*, 012050. [CrossRef]

minerals

Article

Gemological Characteristics of Lvwen Stone and Its Color Genesis

Zhendong Liu [1], Wenjie Wang [2], Ke Yin [1,*], Hanlie Hong [1], Thomas J. Algeo [3,4,5], Zuowei Yin [2], Yong Pan [4], Zhuo Lu [6], Wen Han [7], Yiming Wang [1] and Yunqi Yang [2]

1 School of Earth Sciences, China University of Geosciences, Wuhan 430074, China
2 Gemmological Institute, China University of Geosciences, Wuhan 430074, China
3 State Key Laboratory of Biogeology and Environmental Geology, China University of Geosciences, Wuhan 430074, China
4 State Key Laboratory of Geological Processes and Mineral Resources, China University of Geosciences, Wuhan 430074, China
5 Department of Geosciences, University of Cincinnati, Cincinnati, OH 45221-0013, USA
6 School of Jewelry, West Yunnan University of Applied Sciences, Tengchong 679100, China
7 National Gem & Jewelry Technology Administrative Center, Beijing 100013, China
* Correspondence: yinke@cug.edu.cn

Abstract: "Lvwen stone" is a yellow-green carbonate jade gemstone. In this study, the gemological characteristics and color genesis of Lvwen stone were investigated using conventional gemological testing methods and analytical techniques, such as X-ray diffraction (XRD), Fourier-transform infrared spectroscopy (FTIR), ultraviolet–visible spectroscopy (UV–VIS), laser ablation plasma mass spectrometry (LA-ICP-MS), and scanning electron microscopy (SEM). The chemical composition of Lvwen stone is mainly Ca, with lesser amounts of Mg, Mn, Cu, Zn, Fe, and other trace elements. The rare earth element distribution pattern indicates that Lvwen stone is characterized by MREE depletion and a positive Ce anomaly. The mineralogic composition of Lvwen stone is calcite, and trace-element- and crystal-size-induced colors result in its characteristic banded appearance. The white (or light green) bands consist of comparatively coarse calcite crystals (~100 μm) that are oriented perpendicularly to the band plane, accounting for their poor light transmittance. In contrast, the dark green matrix is composed of cryptocrystalline calcite crystals that are uniform in size (~10 μm) and tightly packed, resulting in superior light transmittance. Lvwen stone has a $^6A_1 \rightarrow ^4E(^4D)$d-d intra-ion electronic transition absorption band of Fe^{3+} at ~380–450 nm and a $^2E \rightarrow ^2T_2(^2D)$d-d intra-ion electronic transition absorption band of Cu^{2+} at ~580–780 nm. This indicates that both the intra-ion electronic transitions of Fe^{3+} and Cu^{2+} give rise to the unique yellow-green color of the material. Lvwen stone is produced by ultra-high-pressure tectonic fluids in a relatively closed, reducing environment, and the green matrix was formed earlier than the white bands.

Keywords: calcite; crystallinity; mineralogy; electronic transition; hydrothermal; trace elements

Citation: Liu, Z.; Wang, W.; Yin, K.; Hong, H.; Algeo, T.J.; Yin, Z.; Pan, Y.; Lu, Z.; Han, W.; Wang, Y.; et al. Gemological Characteristics of Lvwen Stone and Its Color Genesis. *Minerals* **2022**, *12*, 1584. https://doi.org/10.3390/min12121584

Academic Editors: Lidong Dai and Frederick Lin Sutherland

Received: 24 October 2022
Accepted: 4 December 2022
Published: 10 December 2022

1. Introduction

Lvwen stone is a yellow-green gemstone, well-known for its banded appearance of alternating green and white layers. It is a type of carbonate jade, otherwise known as "Afghanistan Jade" (a generic term for carbonate jade) in China. According to its color, structure, transparency, and mineral composition, it is a semiprecious gemstone of the same kind as similar specimens from Pakistan. At present, this jade is known to be produced only in Pakistan [1]. Although Lvwen stone is a gem with a beautiful appearance, it is not common in the jewelry market due to its low hardness.

The nature and origin of Lvwen stones have received only limited attention to date. Hyrsl and Jaroslav [1] reported on its color and mineral composition, describing it as a type of green calcite exhibiting "cat's-eye effect". However, the coloring mechanism, genesis of the cat's-eye effect, and chemical composition of Lvwen stone remain enigmatic. In this study, X-ray diffraction (XRD), Fourier-transform infrared spectroscopy (FTIR), ultraviolet–visible spectroscopy (UV–VIS), laser ablation plasma mass spectrometry (LA-ICP-MS), and scanning electron microscopy (SEM) were used for a comprehensive investigation of typical yellow-green Lvwen stone specimens purchased from the Donghai Jewelry Market, Lianyungang city, China. The goals of the present study are (1) to provide insights into the general mineralogy, chemical composition, microstructural characteristics, and spectroscopic features of Lvwen stones, and (2) to determine the origin of color banding, the mechanism of the cat's-eye effect, and the geological genesis of Lvwen stones.

2. Materials and Methods

2.1. Materials

The three specimens used in this study (S_1, S_2, and S_3) are cabochons, that is, polished but unfaceted gems, with dimensions of ~2 cm (length) × ~1.5 cm (width) × ~0.5 cm (thickness) (Figure 1). Each consists of a dark green matrix with a cryptocrystalline structure, transected by broad, evenly spaced white (or light green) bands that compose the larger part (~60%–70%) of the specimen by volume. The refractive index of all specimens is 1.63. They emit full brightness under a polarizing microscope. These properties indicate that Lvwen stone is a kind of calcite crystal aggregate instead of a single crystal. According to the national standard of the People's Republic of China (GBT16552-2010), natural gems are divided into three categories: natural gemstones (mineral monocrystals or twin crystals), natural jades (mineral assemblages or amorphous matters), and natural organic substances. Because Lvwen stone is a calcite aggregate, it is assigned to the category of natural jades. By using the net water weighing method, the specific gravity of Lvwen stones was determined to be 2.72. They also have an estimated Mohs hardness of about three (3), which is typical of calcite.

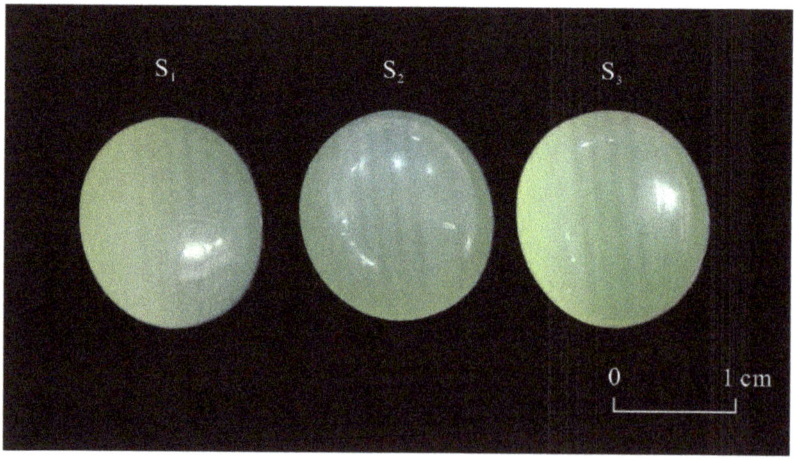

Figure 1. Cabochon specimens (S1, S2, and S3) of yellow-green Lvwen stone. All are top views.

After a cabochon specimen (S_1) was cut into a rectangular prism, it was found that the white bands could still be observed (Figure 2). These white bands are regularly distributed, possibly caused by the lack of chromogenic elements, indicating that the banded structure is not produced by the cat's-eye effect (i.e., chatoyancy, which is due to either a fibrous mineral structure or fibrous inclusions in a mineral). When observed at a low magnification (50×), the green matrix was determined to be microcrystalline, composed of only faintly visible small crystals, whereas the white bands are composed of relatively coarser crystals (~100 μm). The parallel needle-like crystals of the latter are oriented perpendicular to the band planes (Figure 3).

Figure 2. Rectangular slice through cabochon specimen S_1 of Lvwen stone.

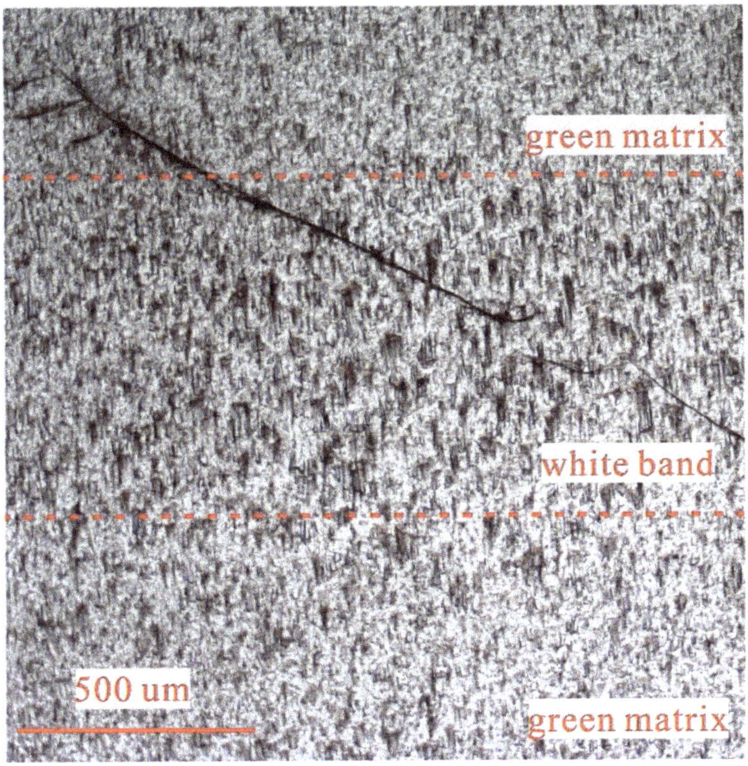

Figure 3. Microscopic image of cabochon specimen S_1 of Lvwen stone under 50× magnification.

2.2. Specimen Preparation and Testing Methods

2.2.1. Specimen Preparation

Fourier-transform infrared spectroscopy (FTIR), ultraviolet–visible spectroscopy (UV–VIS), and laser ablation plasma mass spectrometry (LA-ICP-MS) were conducted by placing the cabochon specimens of Lvwen stone directly on the analytical stage of these instruments. The specimens were then gold-coated and analyzed by scanning electron microscopy (SEM). Furthermore, a portion of the specimen S_1 was ground to powder in an agate mortar to a 200-mesh fineness. The powdered sample was then compacted in a glass holder for X-ray diffraction (XRD) testing.

2.2.2. X-ray Diffraction Analysis

A Panalytical X'Pert PRO DY2198 diffractometer at the State Key Laboratory of Geological Processes and Mineral Resources, China University of Geosciences (Wuhan, China), was utilized to analyze the specimen. The X-ray diffraction (XRD) patterns were recorded with Ni-filtered Cu-Kα radiation (40 kV, 40 mA) and the following testing environment: a temperature of 20 °C and 65% RH with the pattern being recorded from 3° to 64° (2Θ) at a scan rate of 8°/min and a step size of 0.02.

2.2.3. Scanning Electron Microscope

A Quanta 200 environmental scanning electron microscope (SEM) at the State Key Laboratory of Geological Processes and Mineral Resources, China University of Geosciences (Wuhan, China), was used for SEM analysis of the specimens. This analysis was performed under the following conditions: secondary electron imaging, accelerating voltage of 20 kV, and beam current size of ~1–2 nA.

2.2.4. Fourier-Transform Infrared Spectroscopy

Fourier-transform infrared spectroscopy (FTIR) analysis of the Lvwen stone specimens was carried out on an IS50 FTIR spectrometer using the reflection method (equipped with a PIKE Technologies UpIR diffuse reflection accessory) at the Faculty of Materials Science and Chemistry, China University of Geosciences (Wuhan, China). The reflection spectrum was further corrected using the Kramers–Kronig transformation function of OPUS 7.5 to eliminate the effect of optical dispersion on the reflection spectrum. The operating conditions of the FTIR analysis were as follows: 32 scans/min, 4 cm^{-1} resolution, and ~400–4000 cm^{-1} scan range. A thorough washing of the specimens with alcohol before the IR analysis was undertaken to exclude contaminants from prior handling.

2.2.5. Laser Ablation Inductively Coupled Plasma Mass Spectrometry

An Agilent 7700e laser ablation inductively coupled plasma mass spectrometer (LA-ICP-MS) at the Wuhan Sample Solution Analytical Technology Co. Ltd., Wuhan, China, was used to measure the major and trace element concentrations of the Lvwen stone specimens. Detailed operating conditions for the laser ablation system and the ICP-MS instrument and data reduction were the same as described by Zong et al. [2]. A GeoLasPro laser ablation system that consists of a COMPexPro 102 ArF excimer laser (wavelength of 193 nm and maximum energy of 200 mJ, (Agilent Technologies Inc., Santa Clara, CA, USA) and a MicroLas optical system was utilized to perform laser sampling, whereas an Agilent 7700e ICP-MS instrument was used to acquire ion-signal intensities. Helium was applied as a carrier gas, whereas argon was used as the make-up gas and mixed with the carrier gas via a T-connector before entering the ICP-MS. A wire signal smoothing device was included in this laser ablation system [3]. The spot size and frequency of the laser were set to 32 μm and 8 Hz, respectively. The primary standard for major and trace element calibrations was SRM610, and the secondary standards were four USGS materials (MACS-3, BHVO-2G, BCR-2G, and BIR-1G). The latter were treated as unknowns [4]. Each analysis incorporated a background acquisition of approximately 20–30 s, followed by 50 s of data acquisition from the specimens. Off-line selection and integration of background and analyzed signals, time-drift correction, and quantitative calibration for major and trace element analyses were performed by an Excel-based software (ICPMSDataCal 12.2) [4].

2.2.6. Ultraviolet–Visible Spectroscopy

A UV-5000 ultraviolet–visible (UV–VIS) spectrophotometer at the Gemological Institute of the China University of Geosciences (Wuhan, China) was used to undertake UV–VIS absorption spectroscopy. To test the Lvwen stone specimens, the reflection method was applied over a test range of ~100–1200 nm using a scan speed of 800 nm/min and a resolution of 1 nm.

3. Results

3.1. X-ray Diffraction Analysis

The X-ray diffraction analysis of S_1 shows that the main diffraction peaks of Lvwen stone are consistent with those of typical calcite ($CaCO_3$) (PDF:05-0586) (Figure 4). In the ICDD (International Centre for Diffraction Data) reference database, characteristic peaks of 3.84 Å (012), 3.02 Å (104), 2.49 Å (110), 2.28 Å (113), 2.09 Å (202), and 1.91 Å (018) show that calcite is the mineral composing the study specimen. Calculations with the MDI JADE 6.5 software indicate that the cell parameters of this calcite are a = 4.97 Å, c = 17.00 Å, compared with a = 4.99 Å, c = 17.06 Å for typical calcite (PDF:05-0586) *(from ICDD), and that the average size of the calcite crystals is 1737 Å (R = 9.15%).

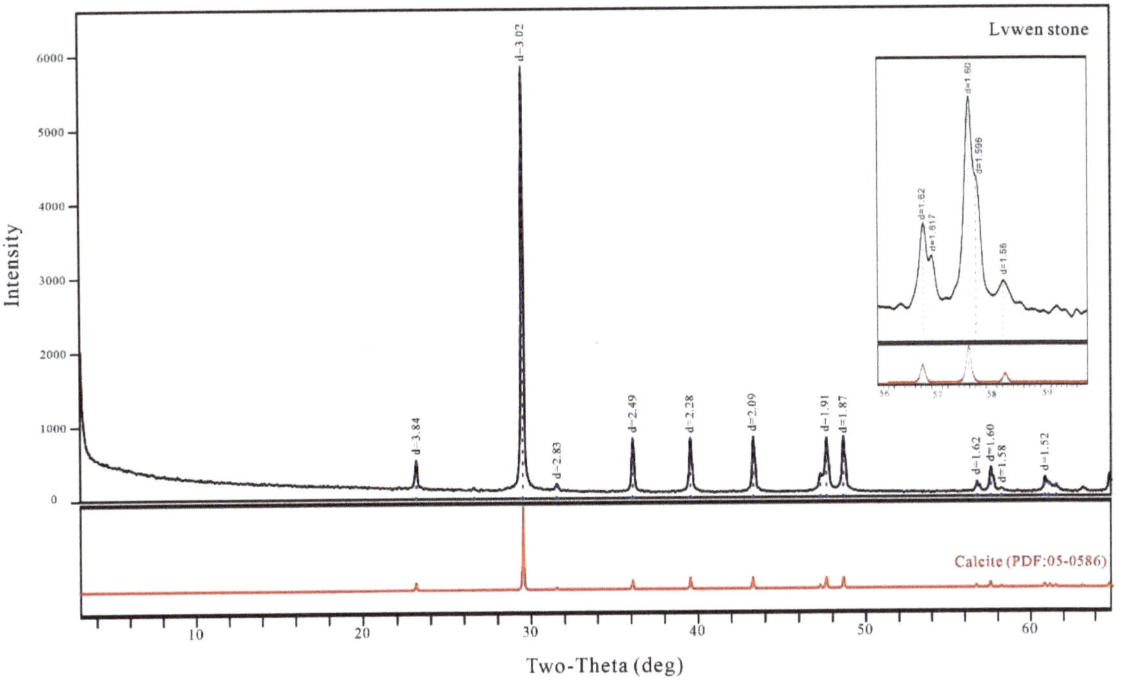

Figure 4. X-ray diffraction pattern of cabochon specimen S_1 of Lvwen stone.

3.2. Spectroscopic Characterization

3.2.1. Fourier-Transform Infrared Spectroscopy Analysis

The Lvwen stone specimens show obvious absorption peaks around 715, 885, ~1125, ~1490, 1796, 2510, 2854, and 2923 cm^{-1} (Figure 5), which are in good agreement with the infrared absorption characteristics of typical calcite [5]. The absorption peaks at 715 and 885 cm^{-1} are especially sharp due to the in-plane and out-of-plane bending vibrations of $[CO_3]^{2-}$, respectively. Broad absorption bands at ~1125 and ~1490 cm^{-1} are linked to the inverse symmetric stretching vibration and inverse asymmetric stretching vibration of $[CO_3]^{2-}$, respectively. Additionally, these two absorption features have similar intensities.

The absorption at 1796 cm^{-1} is assigned to the symmetric stretching vibration and in-plane bending vibration of $[CO_3]^{2-}$ [6]. The double absorptions at 2854 and 2923 cm^{-1} are probably related to organic inclusions of aliphatic hydrocarbons ($-CH_3$ and $-CH_{2-}$) [7].

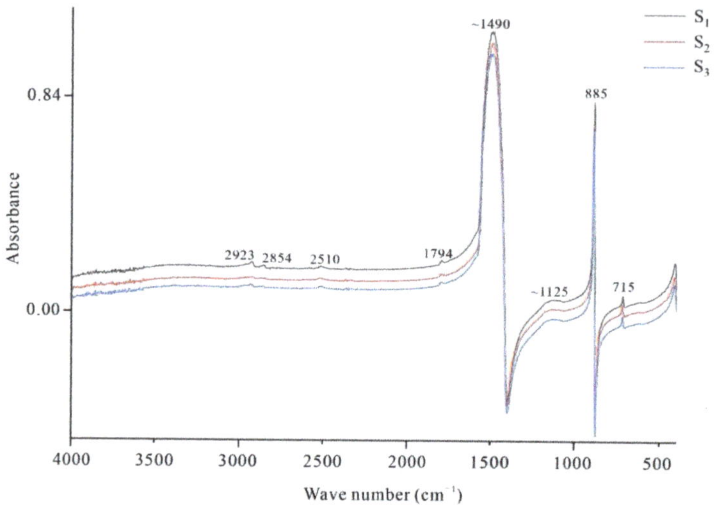

Figure 5. Infrared spectra of three cabochon specimens of Lvwen stone.

3.2.2. Ultraviolet–Visible Spectroscopy Analysis

Two wide absorption bands at ~380–450 nm and ~580–780 nm are displayed in the UV–VIS spectrum (Figure 6). The absorption band at ~380–450 nm is attributed to the $^6A_1 \rightarrow ^4E(^4D)$ d-d electron transition of Fe^{3+}, whereas the absorption band at ~580–780 nm is caused by the $^2E \rightarrow ^2T_2(^2D)$d-d electron transition of Cu^{2+} [8,9]. The absorption band at ~800–1100 nm indicates the presence of Fe^{2+} [10], but it is beyond the range of visible light and does not contribute to the color of Lvwen stone. Therefore, the combined adsorptions of Cu^{2+} and Fe^{3+} in the calcite lattice may account for the yellow-green color of the cabochon specimens of Lvwen stone analyzed in this study.

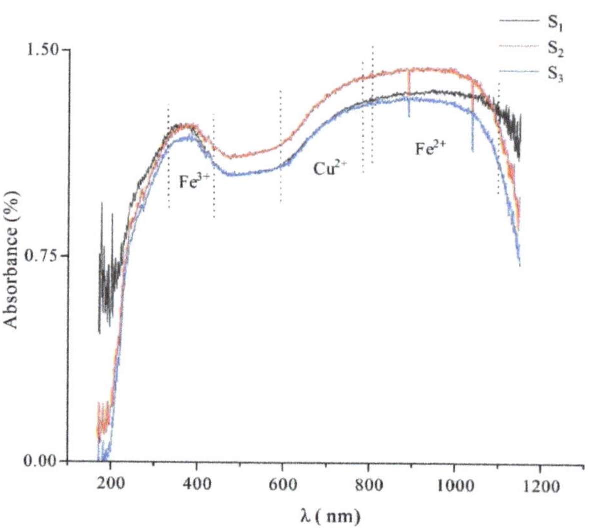

Figure 6. Ultraviolet–visible light (UV–VIS) spectra of three cabochon specimens of Lvwen stone.

3.3. Chemical Compositions

Laser ablation inductively coupled plasma mass spectrometry (LA-ICP-MS) revealed a similar chemical composition for both the green matrix and the white bands of the Lvwen stone specimens—dominantly CaO (>52 wt%), with lesser amounts of MgO and MnO (~0.5 wt% and 1–2 wt%, respectively). Small amounts of Na, Si, K, Sr, and Cu are also present, with the green matrix containing more Cu and Mn than the white bands (Table 1). A number of trace metals were found in the green matrix and white bands, with the amounts of Fe and Zn generally being higher than those of V, Cr, Co, and Ni (Tables 1 and 2). The green matrix exhibits higher concentrations of Fe, Cu, and Zn than the white bands. Because Zn is generally not a coloring agent in minerals, we infer that the yellow-green color of Lvwen stone is linked to the presence of Fe and Cu. The chemical formula of the study specimens was calculated using the anionic method (Table 3), yielding $Ca_{0.97}Mn_{0.02}Mg_{0.01}$ $[CO_3]$.

Table 1. Major element oxide content of three Lvwen stone specimens (all values in wt%).

	NaO	MgO	SiO_2	CaO	MnO	CuO	SrO
S_1-green-1	0.38	0.49	0.21	52.56	1.62	0.32	0.36
S_1-green-2	0.40	0.47	0.21	52.45	1.65	0.34	0.36
S_1-green-3	0.38	0.47	0.26	52.43	1.68	0.35	0.36
Average	0.39	0.48	0.24	52.48	1.65	0.33	0.36
S_2-green-1	0.36	0.51	0.30	52.44	1.63	0.30	0.36
S_2-green-2	0.35	0.47	0.31	52.29	1.77	0.36	0.35
S_2-green-3	0.35	0.47	0.30	52.38	1.72	0.34	0.36
Average	0.35	0.48	0.30	52.37	1.71	0.33	0.36
S_3-green-1	0.35	0.49	0.31	52.57	1.54	0.29	0.36
S_3-green-2	0.36	0.50	0.31	52.44	1.62	0.28	0.37
S_3-green-3	0.35	0.48	0.31	52.33	1.73	0.35	0.35
Average	0.35	0.49	0.31	52.45	1.63	0.31	0.36
S_1-white-1	0.38	0.56	0.21	53.00	1.16	0.19	0.40
S_1-white-2	0.40	0.57	0.21	52.84	1.27	0.21	0.39
S_1-white-3	0.42	0.58	0.26	52.85	1.19	0.18	0.39
Average	0.40	0.57	0.24	52.90	1.20	0.19	0.39
S_2-white-1	0.36	0.56	0.30	53.02	1.14	0.17	0.38
S_2-white-2	0.37	0.59	0.30	52.98	1.17	0.17	0.37
S_2-white-3	0.37	0.59	0.32	52.92	1.17	0.17	0.38
Average	0.37	0.58	0.31	52.98	1.16	0.17	0.38
S_3-white-1	0.37	0.57	0.32	52.96	1.18	0.17	0.37
S_3-white-2	0.36	0.57	0.31	53.04	1.14	0.16	0.37
S_3-white-3	0.36	0.57	0.33	53.11	1.08	0.15	0.36
Average	0.36	0.57	0.32	53.04	1.14	0.15	0.37

S_1-green-1 and S_1-white-1 refer to the random first point of the green matrix and white bands, respectively, of S_1, and so on.

Table 2. Trace element content of three Lvwen stone specimens (all values in ppm).

	V	Cr	Fe	Co	Ni	Zn
S_1-green-1	0.34	0.51	185.88	38.88	3.20	399.46
S_1-green-2	0.26	0.76	204.03	38.36	3.96	416.37
S_1-green-3	0.28	0.78	237.38	39.29	3.22	453.73
Average	0.29	0.68	209.10	38.84	3.46	423.19
S_2-green-1	0.18	0.46	218.61	36.80	2.98	407.60
S_2-green-2	0.24	0.12	268.16	36.89	3.70	471.20
S_2-green-3	0.29	0.00	233.86	36.55	1.71	400.73
Average	0.24	0.20	240.21	36.75	2.80	426.51

Table 2. *Cont.*

	V	Cr	Fe	Co	Ni	Zn
S_3-green-1	0.31	0.00	223.69	34.95	2.30	382.36
S_3-green-2	0.29	0.00	231.72	37.30	3.44	375.34
S_3-green-3	0.28	0.00	276.33	36.38	1.80	443.82
Average	0.29	0.00	243.91	37.30	2.51	400.51
S_1-white-1	0.21	0.74	135.69	34.99	3.47	226.68
S_1-white-2	0.30	1.40	151.05	35.70	2.63	251.09
S_1-white-3	0.25	0.46	109.16	35.07	3.62	220.96
Average	0.25	0.87	131.97	35.25	3.24	232.91
S_2-white-1	0.26	0.99	165.77	36.05	3.23	211.17
S_2-white-2	0.17	0.74	112.69	37.96	2.95	232.60
S_2-white-3	0.18	0.00	137.22	38.77	3.07	242.52
Average	0.20	0.58	138.56	37.59	3.08	228.76
S_3-white-1	0.29	1.28	167.06	36.46	2.29	243.10
S_3-white-2	0.27	0.00	124.00	36.97	2.78	241.07
S_3-white-3	0.25	0.00	117.14	36.46	1.54	255.99
Average	0.27	0.43	136.07	36.63	2.20	246.72

S_1-green-1 and S_1-white-1 refer to the random first point of the green matrix and white bands, respectively, of S_1, and so on.

Table 3. Molar composition of three Lvwen stone specimens.

	The Number of Cations Based on Of.u. = 3							Stoichiometric Number
	S_1-green	S_2-green	S_3-green	S_1-white	S_2-white	S_3-white	Total Average	
Mg^{2+}	0.01	0.01	0.01	0.01	0.01	0.01	0.01	
Ca^{2+}	0.96	0.96	0.96	0.97	0.97	0.97	0.97	1.00
Mn^{2+}	0.02	0.02	0.02	0.02	0.02	0.02	0.02	
C^{4+}	1.00	1.00	1.00	1.00	1.00	1.00	1.00	1.00

S_1-green and S_1-white refer to the average chemical compositions of the green matrix and white bands, respectively, of S_1, and so on.

The total amount of rare earth elements (ΣREE) is ~14.3–36.0 ppm, with the green matrix being significantly enriched (28.4–36.0 ppm; mean 30.8 ppm) relative to the white bands (14.3–23.9 ppm; mean 19.6 ppm). Both the green matrix and the white bands display a pattern of LREE and HREE enrichment and MREE depletion (Table 4, Figure 7). Cerium anomalies (δCe, calculated as $\delta Ce = [Ce]/\sqrt{[La] \times [Pr]}$, where [Ce], [La], and [Pr] are chondrite-normalized values) are positive for both the green matrix (1.18–1.29) and white bands (1.04–1.25), with an overall mean of 1.21. Europium anomalies (δEu, calculated as $\delta Eu = [Eu]/\sqrt{[Sm] \times [Gd]}$, where [Eu], [Sm], and [Gd] are chondrite-normalized values) are negative for both the green matrix (0.69–1.14) and white bands (0.65–1.02), with an overall mean of 0.84 (Table 4, Figure 7).

Table 4. REE characteristics of three Lvwen stone specimens (all values in ppm).

	S_1-green-1	S_1-green-2	S_1-green-3	S_2-green-1	S_2-green-2	S_2-green-3	S_3-green-1	S_3-green-2	S_3-green-3	Average
La	3.36	3.64	4.14	3.33	3.74	3.16	3.27	3.48	3.68	3.53
Ce	8.20	9.23	10.40	7.84	8.92	7.87	8.64	8.32	9.04	8.72
Pr	0.81	0.94	0.94	0.77	0.83	0.80	0.90	0.81	0.78	0.84
Nd	4.05	4.66	5.06	4.15	4.42	3.80	4.29	3.83	4.34	4.29
Sm	1.00	1.25	1.31	0.97	1.31	1.24	1.18	1.02	1.51	1.20
Eu	0.38	0.51	0.50	0.51	0.48	0.38	0.43	0.40	0.48	0.45
Gd	1.95	2.14	2.67	1.90	2.02	2.29	1.99	2.29	2.06	2.14
Tb	0.39	0.40	0.44	0.37	0.40	0.37	0.42	0.37	0.41	0.40

Table 4. *Cont.*

	S_1-green-1	S_1-green-2	S_1-green-3	S_2-green-1	S_2-green-2	S_2-green-3	S_3-green-1	S_3-green-2	S_3-green-3	Average	
Dy	2.96	3.30	3.41	2.68	3.11	2.90	2.42	2.50	3.04	2.92	
Ho	0.79	0.82	0.94	0.68	0.76	0.74	0.77	0.69	0.78	0.77	
Er	2.46	2.62	2.71	2.29	2.36	2.12	2.38	2.08	2.65	2.41	
Tm	0.35	0.35	0.38	0.32	0.32	0.35	0.33	0.36	0.33	0.34	
Yb	2.46	2.50	2.72	2.27	2.54	2.49	2.11	2.26	2.35	2.41	
Lu	0.33	0.30	0.37	0.33	0.37	0.32	0.33	0.26	0.35	0.33	
ΣREE	29.49	32.66	35.98	28.42	31.57	28.83	29.46	28.67	31.80	30.76	
δCe	1.20	1.21	1.27	1.18	1.23	1.19	1.22	1.20	1.29		
δEu	0.84	0.95	0.81	1.14	0.90	0.69	0.85	0.81	0.83		
	S_1-white-1	S_1-white-2	S_1-white-3	S_2-white-1	S_2-white-2	S_2-white-3	S_3-white-1	S_3-white-2	S_3-white-3	Average	Chondrite *
La	2.03	2.24	1.72	2.46	2.26	2.64	2.36	1.97	2.00	2.19	0.24
Ce	4.93	5.59	4.05	5.87	5.65	6.64	6.14	5.21	5.39	5.50	0.61
Pr	0.45	0.58	0.38	0.57	0.57	0.63	0.61	0.74	0.66	0.58	0.09
Nd	2.41	2.63	1.80	2.74	2.76	3.42	2.99	2.45	2.53	2.64	0.46
Sm	0.60	0.81	0.65	1.00	0.80	1.01	0.90	0.88	0.97	0.85	0.15
Eu	0.28	0.24	0.21	0.27	0.31	0.34	0.25	0.33	0.30	0.28	0.06
Gd	1.19	1.27	0.86	1.46	1.28	1.65	1.51	1.48	1.44	1.35	0.20
Tb	0.20	0.25	0.19	0.24	0.24	0.24	0.23	0.35	0.32	0.25	0.04
Dy	1.73	1.70	1.55	2.06	1.91	2.49	2.18	2.14	2.06	1.98	0.25
Ho	0.41	0.50	0.35	0.49	0.52	0.58	0.59	0.54	0.50	0.50	0.05
Er	1.33	1.58	1.09	1.68	1.56	1.86	1.68	1.47	1.06	1.48	0.16
Tm	0.20	0.22	0.17	0.27	0.24	0.27	0.30	0.25	0.28	0.24	0.02
Yb	1.26	1.49	1.13	1.75	1.63	1.83	1.75	1.58	1.63	1.56	0.16
Lu	0.23	0.21	0.17	0.26	0.29	0.25	0.25	0.26	0.28	0.25	0.02
ΣREE	17.24	19.31	14.32	21.11	20.01	23.87	21.73	19.64	19.42	19.63	
δCe	1.24	1.18	1.22	1.20	1.20	1.25	1.24	1.04	1.13		
δEu	1.02	0.72	0.86	0.68	0.94	0.81	0.65	0.89	0.77		

S_1-green and S_1-white refer to the average chemical compositions of THE green matrix and white bands, respectively, of S_1, and so on. * The REE data of chondrites are from [11]. McDonough WF, Sun SS. The composition of the Earth. Chem Geol. 1995, 120, 223–253.

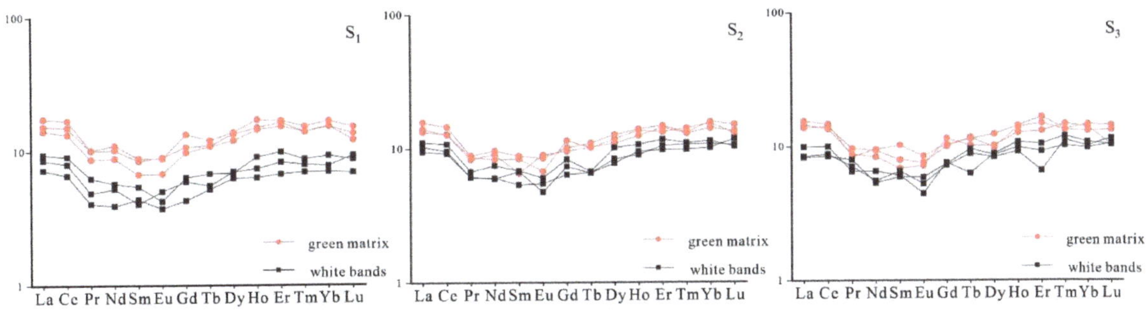

Figure 7. Chondrite-normalized REE patterns of three cabochon specimens of Lvwen stone.

3.4. Microstructural Characteristics

Scanning electron microscopy (SEM) revealed that most of the calcite crystals in the white bands of Lvwen stone (S_1) are in the form of crystal aggregates with a size of 30−100 um. Moreover, the crystal interstices are filled with nanoscale particles and cryptocrystalline colloids. The dominant orientation of these crystals is perpendicular to the plane of the white bands (Figure 8a,b). Composed of mainly cryptocrystalline calcite with trace nanoscale particles filling in the intergranular spaces, the calcite crystals in the green matrix are not oriented and are mainly ~10 μm in size (Figure 8c,d). The crystallinity of calcite in the white bands is significantly higher than that of the green matrix. These findings are consistent with the results from optical microscopy (Figure 3).

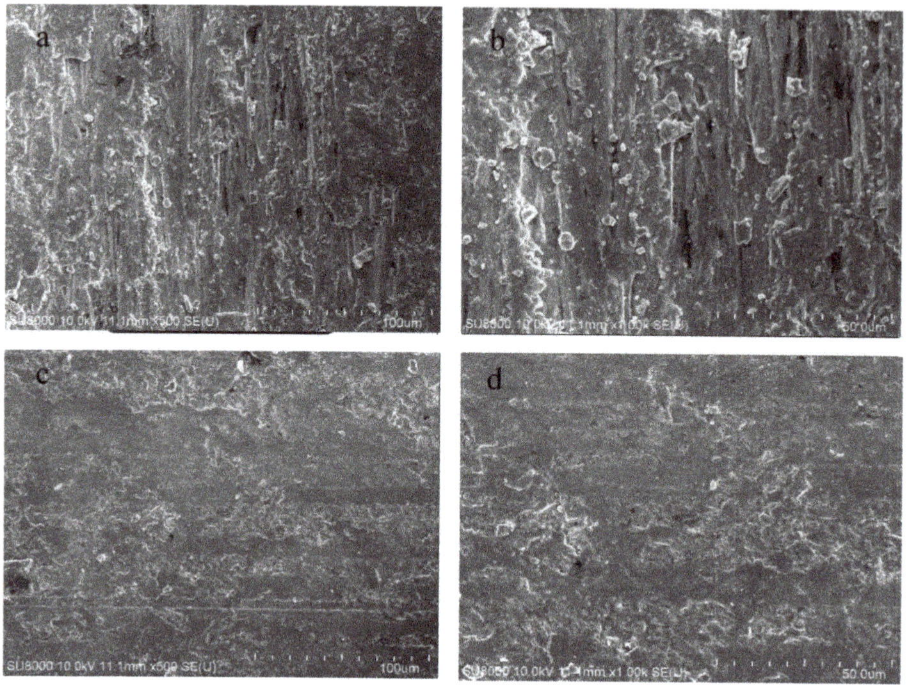

Figure 8. The micromorphology of minerals in cabochon specimen S_1 of Lvwen stone (SEM images). (**a**,**b**) are 500× and 1000× images of white bands, respectively; (**c**,**d**) are 500× and 1000× images of green matrix, respectively.

4. Discussion

The results of the XRD, FTIR, and LA-ICP-MS analyses indicate that the mineral composition of the Lvwen stone specimens is calcite with a chemical formula of $Ca_{0.97}Mn_{0.02}Mg_{0.01}$ $[CO_3]$. The yellow-green hue is probably due to the presence of Fe^{3+} and Cu^{2+}. Based on cell parameter calculations, the dimensions of the a_0, b_0, and c_0 axes of calcite generally decrease with increasing isomorphic substitution of Mg^{2+} and Mn^{2+} for Ca^{2+} in the crystal lattice [12]. Therefore, isomorphic substitution by Mg^{2+} and Mn^{2+} may be the cause of the relatively low a_0, b_0, and C_0 values of calcite in the Lvwen stone specimens. Isomorphic substitution may also account for the appearance of the ~1125 cm^{-1} absorption band and the broadening of the ~1490 cm^{-1} band in FTIR spectra. Thus, this process may be the main factor in the reduced symmetry of the $[CO_3]^{2-}$ distribution in the Lvwen stone specimens compared with that of ideal calcite [5]. Two diffraction peaks (1.617 and 1.596 Å) in the XRD spectrum (Figure 4), which appear to the right of the 1.62 and 1.60 Å peaks of typical calcite (PDF:05-0586), may be due to the reduced symmetry of $[CO_3]^{2-}$ in the calcite crystals, as revealed by FTIR.

To better study the banded structure of Lvwen stone, the S_1 cabochon was cut down into a rectangular prism. The development of banding is linked to variations in the concentrations of trace-metal ions: the green matrix has a higher content of ΣREE, Cu, and Fe than the white bands (Tables 2 and 4). Therefore, the green matrix of Lvwen stone may have formed when REEs, Cu, Fe, and other trace elements were abundant in the mineralizing fluid, whereas the white bands formed after these elements had been depleted. Micrograph and SEM observations show that the white bands contain mainly micron-sized, oriented calcite crystals, whereas the green matrix is composed of cryptocrystalline calcite and smaller crystals than the white bands (Figure 8a–d). The white bands, consisting of larger crystals, also harbor larger pores, which are the cause of their poorer light transmission. In contrast, the green matrix is composed of smaller crystals and has correspondingly smaller pores, giving rise to better light transmission. In addition, the different crystal particle sizes may also lead to the different color of the green matrix and the white bands. However, there is no research on the relationship between crystal sizes and color in natural minerals. LA-ICP-MS analysis showed that the green matrix has significantly higher contents of Fe^{3+} and Cu^{2+} than the white bands (Tables 1 and 2). Furthermore, broad absorptions are displayed at ~380–450 and ~580–780 nm in the UV–VIS spectrum, which are related to the $^6A_1 \rightarrow ^4E(^4D)$d-d intra-ion electron transition of Fe^{3+} and $^2E \rightarrow ^2T_2(^2D)$d-d intra-ion electron transition of Cu^{2+}, respectively [8,9]. Absorptions in the green (~490–570 nm) and yellow (~570–580 nm) regions of the UV–VIS spectrum are relatively weak, and the mixture of these two colors produces the yellow-green color of Lvwen stone.

The green matrix and the white bands in Lvwen stone have a shared origin, that is, being derived from a single fluid, as inferred from a Y/Ho-La/Ho discriminant plot in which all samples yield nearly the same values (Figure 9) [13]. A hydrothermal origin can be inferred from a Tb/Ca-Tb/La discriminant plot, which shows $\log_{10}$(Tb/Ca) of −6.53 to −6.90 and $\log_{10}$(Tb/La) of −0.99 to −1.06 (Figure 10) [14]. Calcite tends to take up REEs readily owing to isomorphic substitution for Ca^{2+}, which has a similar ironic radius [15]. Hydrothermal fluids tend to yield REE distribution patterns characterized by (1) LREE enrichment and HREE depletion, or (2) LREE and HREE depletion, and MREE enrichment [16,17]. However, Lvwen stone displays a different pattern, characterized by LREE and HREE enrichment and MREE depletion, suggesting derivation from MREE-deficient mineralizing fluids in a closed system. Among the REEs, Ce is a variable-valence element that is sensitive to changes in its redox environment. In a reducing environment, Ce is present as Ce^{3+}, which is not fractionated relative to other trivalent REEs, but in an oxidizing environment, Ce^{4+} is precipitated as the highly insoluble mineral cerianite [CeO_2], resulting in a negative Ce anomaly in the residual fluid [15,18]. We infer that the positive Ce anomalies (Table 4 and Figure 7) shown by both the green matrix and white bands indicate the reductive remobilization of earlier precipitated Ce-oxides by the mineralizing fluid that yielded Lvwen stone. The negative Eu anomalies (Table 4 and Figure 7) shown by both the green matrix and white bands are considered to be indicative of reducing conditions [13], which is consistent with the reductive mineralizing fluid revealed by the positive Ce anomalies. The calcite crystals in the white bands are oriented, suggesting that they formed in a specific tectonic stress regime. Previous investigations have shown that low-permeability rocks (e.g., mudstones) in foreland basins or orogenic zones have the tendency to form fibrous and parallel-growing calcite veins under ultra-high-pressure tectonic fluids, with the growth direction of the calcite crystals being perpendicular (or nearly so) to the plane of the vein [19–21]. We infer that Lvwen stone may have formed from such a tectonically controlled mineralizing fluid. This hypothesis is supported by the presence of fatty hydrocarbon inclusions in the calcite of the white bands, as revealed by adsorption peaks at 2923 and 2854 cm^{-1} in the infrared spectrum [22]. Hydrocarbon inclusions, which record the formation of fluids under abnormal pressure during tectonic deformation, are often found in calcite veins [22].

Isostructural veins in calcite crystals are divisible into three kinds based on crystal structure, morphology, and fiber arrangement: stretching veins, syntaxial veins, and antitaxial veins [23,24]. The white bands of Lvwen stone share characteristics with the stretching vein type. Stretching veins are formed in fractures filled with a vein-forming fluid from the surrounding rocks or early-generated veins, in which the growth of veins proceeds by repeated cycles of opening and closing of the host fracture [25]. During the formation of a stretching vein, a given fracture does not necessarily open at exactly the same location every time, and small changes in fracture position from one cycle to the next results in shifts in the growth interface [26], leading to the uneven boundaries of the stretching vein. Following multiple cycles of opening and closing of a fracture, the remains of the country rock or of earlier veins running parallel to fractures often appear within a vein [23,27,28]. The margins of the white bands of Lvwen stone are typically ragged, indicating that the vein-forming crystals have undergone periodic tectonic stretching [26]. The green matrix of Lvwen stone may have been a cryptocrystalline vein formed at an earlier stage, which then, due to repetitive opening and closing of tectonic fractures, gradually acquired the multiple white bands with parallel crystals seen in the study specimens, thus yielding their present, alternately green- and white-banded appearance.

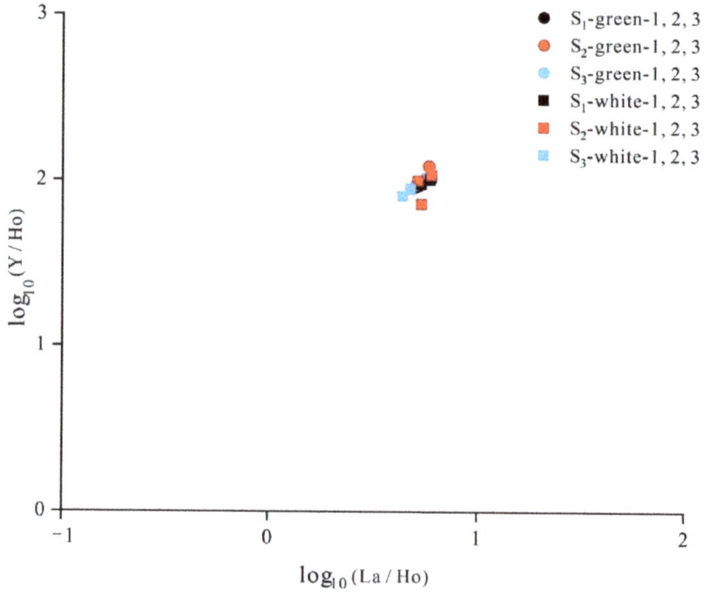

Figure 9. Y/Ho-La/Ho crossplot of three cabochon specimens of Lvwen stone. Values represent atomic ratios rather than weight ratios.

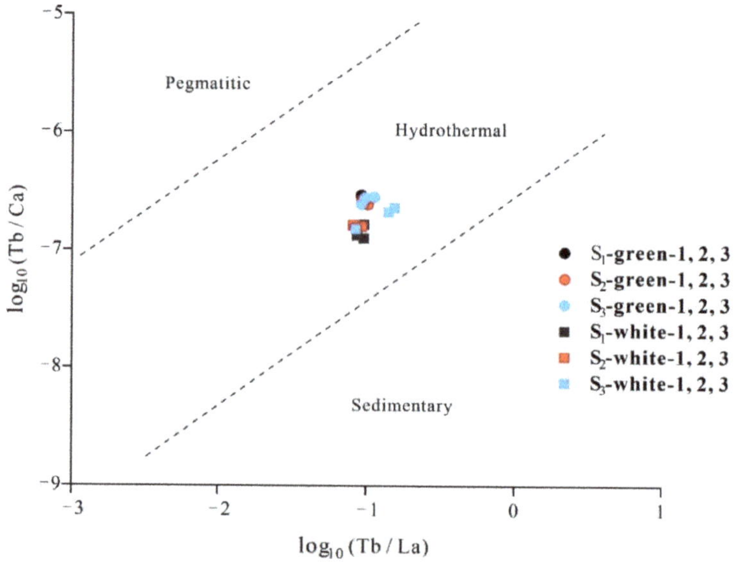

Figure 10. Tb/Ca-Ta/La discriminant plot of three cabochon specimens of Lvwen stone. Values represent atomic ratios rather than weight ratios.

5. Conclusions

(1) The mineral composition of Lvwen stone is calcite, with crystal cell parameters of a = 4.97 and c = 17.00 Å and calcite crystals with an average grain size of 1737 Å. The cabochons used in this study have obvious white bands even after being cut into blocks, indicating that the banding is not due to chatoyancy ("cat's-eye effect"). The white bands are composed of parallel-arranged, ~100 μm–sized calcite crystals. In contrast, the green matrix is dominated by smaller (~10 μm) cryptocrystalline calcite particles. These differences in crystallinity contribute to the contrasting transparencies of the white bands and green matrix.

(2) The chemical formula of Lvwen stone is $Ca_{0.97}Mn_{0.02}Mg_{0.01}$ $[CO_3]$, and the content of ΣREE in the green matrix is higher than that in the white bands. There is a significantly larger content of Fe and Cu in the green matrix as compared with the white bands, showing that the green color of the matrix is formed by the absorption of light by Fe^{3+} and Cu^{2+} in the calcite lattice, resulting from the $^6A_1 \rightarrow ^4E(^4D)$d-d intra-ion electron transition of Fe^{3+} and $^2E \rightarrow ^2T_2(^2D)$d-d intra-ion electron transition of Cu^{2+}, respectively. In addition, different crystal particle sizes may also affect the color of the green matrix and the white bands.

(3) The occurrence of the ~1125 cm^{-1} absorption band and the broadening of the ~1490 cm^{-1} band on the FTIR spectrum suggest that the reduced symmetry of the $[CO_3]^{2-}$ distribution of calcite crystals in Lvwen stone, compared with that of ideal calcite, may be related to isomorphic substitution of Mg^{2+} and Mn^{2+} for Ca^{2+}. The double absorptions at 2923 and 2854 cm^{-1} reveal the presence of aliphatic hydrocarbon inclusions in the study specimens.

(4) Discriminant diagrams show that the green matrix and white bands of Lvwen stone had the same origin, and that both are the products of hydrothermal fluids. The presence of positive Ce anomalies in both the green matrix and white bands is an indication that the mineralizing fluid reductively remobilized earlier-formed Ce-oxides that had precipitated under oxidizing conditions. The orientation of crystals in the white bands perpendicular to the band plane shows that the study specimens formed in a tectonically controlled stress regime that permitted the formation of stretching veins.

Author Contributions: Conceptualization, W.H.; methodology, Z.Y.; software, Y.W.; investigation, W.W.; resources, Y.P.; data curation, Y.Y.; writing—original draft preparation, Z.L. (Zhendong Liu); writing—review and editing, T.J.A. and H.H.; supervision, K.Y.; funding acquisition, Z.L. (Zhuo Lu). All authors have read and agreed to the published version of the manuscript.

Funding: This study was supported by National Natural Science Foundation of China (Nos. 41772045 and 41402036), Fundamental Research Funds for the Central Universities (No. CUG170106), Science and Technology Projects Entrusted by Enterprises and Institutions (2021016719), University Gemstone and Material Process Engineering Research Center Project of Yunnan Province (No. ZX20200103), and Scientific Research Fund Project of Yunnan Education Department (No. 2019J0311).

Data Availability Statement: The authors confirm that the data supporting the findings of this study are available within the article.

Acknowledgments: We are grateful to Dong Wang for the purchase of the Lvwen stone specimens. Our gratitude extends to the Wuhan Sample Solution Analytical Technology Co., Ltd., Wuhan, China, for their technical support during the in situ geochemical compositions of Lvwen stone; Qin Yang for SEM observation; and Jishun Yu for XRD analyses. We also thank three anonymous reviewers and guest editor, Lidong Dai for their constructive comments.

Conflicts of Interest: The authors declare no conflict of interest.

References

1. Hyrsl, J. Cat's-eye calcite from Pakistan. *Gems. Gemol.* **2015**, *51*, 199–200.
2. Zong, K.; Klemd, R.; Yuan, Y.; He, Z.; Guo, J.; Shi, X.; Liu, Y.; Hu, Z.; Zhang, Z. The assembly of Rodinia: The correlation of early Neoproterozoic (ca. 900Ma) high-grade metamorphism and continental arc formation in the southern Beishan Orogen, southern Central Asian Orogenic Belt (CAOB). *Precambrian Res.* **2017**, *290*, 32–48. [CrossRef]
3. Hu, Z.; Zhang, W.; Liu, Y.; Gao, S.; Li, M.; Zong, K.; Chen, H.; Hu, S. "Wave" Signal-Smoothing and Mercury-Removing Device for Laser Ablation Quadrupole and Multiple Collector ICPMS Analysis: Application to Lead Isotope Analysis. *Anal. Chem.* **2015**, *87*, 1152–1157. [CrossRef] [PubMed]
4. Liu, Y.; Hu, Z.; Gao, S.; Günther, D.; Xu, J. In situ analysis of major and trace elements of anhydrous minerals by LA-ICP-MS without applying an internal standard. *Chem. Geol.* **2008**, *257*, 34–43. [CrossRef]
5. Wen, L.; Liang, W.; Zhang, Z.; Huang, J. *The Infrared Spectroscopy of Minerals*; Chongqing University Press: Chongqing, China, 1989.
6. Plav, B.; Kobe, S.; Orel, B. Identification of crystallization forms of $CaCO_3$ with FTIR spectroscopy. *Kovine Zlitine Tehnol.* **1999**, *33*, 517–521.
7. Li, R.; Dong, S.; Lehrmann, D.; Duan, L. Tectonically driven organic fluid migration in the Dabashan Foreland Belt: Evidenced by geochemistry and geothermometry of vein-filling fibrous calcite with organic inclusions. *J. Asian Earth Sci.* **2013**, *75*, 202–212. [CrossRef]
8. Marfunin, A.d.S.; Egorova, N. *Physics of Minerals and Inorganic Materials: An introduction*; Springer: Berlin/Heidelberg, Germany, 1979.
9. Farges, F.; Benzerara, K.; Brown, G.E., Jr. Chrysocolla redefined as spertiniite. *Am. Inst. Phys.* **2007**, *882*, 223–225.
10. Gaffey, S.J. Spectral reflectance of carbonate minerals in the visible and near infrared (0.35–2.55 um): Anhydrous carbonate minerals. *J. Geophys. Res. Solid Earth* **1987**, *92*, 1429–1440. [CrossRef]
11. McDonough, W.F.; Sun, S.S. The composition of the Earth. Chem Geol. *Chem. Geol.* **1995**, *120*, 223–253. [CrossRef]
12. Zhao, S. *Crystallography and Mineralogy (Third Edition)*; Higher Education Press: Beijing, China, 2017.
13. Bau, M.; Möller, P. Rare earth element fractionation in metamorphogenic hydrothermal calcite, magnesite and siderite. *Mineral. Petrol.* **1992**, *45*, 231–246. [CrossRef]
14. Möller, P.; Parekh, P.P.; Schneider, H.J. The application of Tb/Ca-Tb/La abundance ratios to problems of fluorspar genesis. *Miner. Depos.* **1976**, *11*, 111–116. [CrossRef]
15. Zhong, S.; Mucci, A. Partitioning of rare earth elements (REEs) between calcite and seawater solutions at 25 C and 1 atm, and high dissolved REE concentrations. *Geochim. Cosmochim. Acta* **1995**, *59*, 443–453. [CrossRef]
16. Lottermoser, B. Rare earth elements and hydrothermal ore formation processes. *Ore Geol. Rev.* **1992**, *7*, 25–41. [CrossRef]
17. Haas, J.R.; Shock, E.L.; Sassani, D.C. Rare earth elements in hydrothermal systems: Estimates of standard partial molal thermodynamic properties of aqueous complexes of the rare earth elements at high pressures and temperatures. *Geochim. Cosmochim. Acta* **1995**, *59*, 4329–4350. [CrossRef]
18. Migdisov, A.; Williams-Jones, A.; Brugger, J.; Caporuscio, F.A. Hydrothermal transport, deposition, and fractionation of the REE: Experimental data and thermodynamic calculations. *Chem. Geol.* **2016**, *439*, 13–42. [CrossRef]
19. Machel, H.G. Fibrous gypsum and fibrous anhydrite in veins. *Sedimentology* **1985**, *32*, 443–454. [CrossRef]
20. Al-Aasm, I.S.; Muir, I.; Morad, S. Diagenetic conditions of fibrous calcite vein formation in black shales: Petrographic, chemical and isotopic evidence. *Bull. Can. Pet. Geol.* **1993**, *41*, 46–56.
21. Badertscher, N.P.; Beaudoin, G.; Therrien, R.; Burkhard, M. Glarus overthrust: A major pathway for the escape of fluids out of the Alpine orogen. *Geology* **2002**, *30*, 875–878. [CrossRef]
22. Parnell, J.; Honghan, C.; Middleton, D.; Haggan, T.; Carey, P. Significance of fibrous mineral veins in hydrocarbon migration: Fluid inclusion studies. *J. Geochem. Explor.* **2000**, *69*, 623–627. [CrossRef]
23. Bons, P.D.; Montenari, M. The formation of antitaxial calcite veins with well-developed fibres, Oppaminda Creek, South Australia. *J. Struct. Geol.* **2005**, *27*, 231–248. [CrossRef]
24. Bons, P.D.; Elburg, M.A.; Gomez-Rivas, E. A review of the formation of tectonic veins and their microstructures. *J. Struct. Geol.* **2012**, *43*, 33–62. [CrossRef]
25. Wang, M.; Chen, Y.; Xu, X. Progress on formation mechanism of the fibrous veins in mudstone and its implications to hydrocarbon migration. *Adv. Earth Sci.* **2015**, *30*, 1107–1118.
26. Durney, D.W.; Ramsay, J.G. Incremental strains measured by syntectonic crystal growths. In *Gravity and tectonics*; De Jong, K.A., Scholten, R., Eds.; Wiley: New York, NY, USA, 1973; pp. 67–96.
27. Ramsay, J.G. The crack-seal mechanism of rock deformation. *Nature* **1980**, *284*, 135–139. [CrossRef]
28. Hilgers, C.; Urai, J.L. Microstructural observations on natural syntectonic fibrous veins: Implications for the growth process. *Tectonophysics* **2002**, *352*, 257–274. [CrossRef]

MDPI
St. Alban-Anlage 66
4052 Basel
Switzerland
Tel. +41 61 683 77 34
Fax +41 61 302 89 18
www.mdpi.com

Minerals Editorial Office
E-mail: minerals@mdpi.com
www.mdpi.com/journal/minerals

www.ingramcontent.com/pod-product-compliance
Lightning Source LLC
Chambersburg PA
CBHW061621210926
51850CB00009B/2391
9783036572901